An Urban Politics of Climate Change is a *tour de force.* The book effectively blends complex urban political theory with practical, on-the-ground examples of climate action and cutting edge experimentation. The authors are able to capture the current moment of how cities are beginning to respond to the climate change challenge while providing a deeply-rooted, scholarly lens into the implications of these efforts and how climate change action is simultaneously revealing existing tensions within cities and opening up opportunities for progressive forms of governance. The volume will be of interest to all – students, academics, and practitioners alike.

William Solecki, Professor of Geography, Hunter College – City University of New York

By placing at the centre of their study the key but under-explored role of experimentations in the shaping of city-level responses to climate change, Bulkeley, Castán Broto and Edwards open new perspectives on the politics of transformative socio-material change in cities in the age of the anthropocene.

Olivier Coutard, Director, LATTS, Université Paris-Est, CNRS

This is an excellent and thoughtful collection of case studies from around the world, combining urban experimentations with climate change. It is skilfully set in the context of transition theory and the politics of environmental governance.

Professor Simin Davoudi, Institute for Sustainability, Newcastle University, UK

AN URBAN POLITICS OF CLIMATE CHANGE

The confluence of global climate change, growing levels of energy consumption and rapid urbanization has led the international policy community to regard urban responses to climate change as 'an urgent agenda' (World Bank 2010). The contribution of cities to rising levels of greenhouse gas emissions coupled with concerns about the vulnerability of urban places and communities to the impacts of climate change have led to a relatively recent and rapidly proliferating interest amongst both academic and policy communities in how cities might be able to respond to mitigation and adaptation. Attention has focused on the potential for municipal authorities to develop policy and plans that can address these twin issues, and the challenges of capacity, resource and politics that have been encountered. While this literature has captured some of the essential means through which the urban response to climate change is being forged, it has failed to take account of the multiple sites and spaces of climate change response that are emerging in cities 'off-plan'.

An Urban Politics of Climate Change provides the first account of urban responses to climate change that moves beyond the boundary of municipal institutions to critically examine the governing of climate change in the city as a matter of both public and private authority, and to engage with the ways in which this is bound up with the politics and practices of urban infrastructure. Drawing on cases from multiple cities in both developed and emerging economies, this volume provides new insight into both the potential and the limitations of urban responses to climate change, as well as suggesting new conceptual direction for our understanding of the politics of environmental governance.

Harriet Bulkeley is Professor of Geography at the University of Durham. Her research interests are in the nature and politics of environmental governance, with a particular focus on climate change, energy and urban sustainability. She is author of *Climate Change and the City* (Routledge Critical Introductions to the City 2012), and (with Peter Newell) *Governing Climate Change* (Routledge 2010).

Vanesa Castán Broto is a Lecturer at the Development and Planning Unit of the Barlett Faculty of the Built Environment in University College London. She teaches at the post-graduate level in urban political ecology and sustainable development, urban development planning and urban resilience.

Gareth A. S. Edwards is a Lecturer in Geography and Development in the School of International Development at the University of East Anglia. His research interests centre on the ethics of environmental governance, with a focus on the political ecology of climate change and water.

AN URBAN POLITICS OF CLIMATE CHANGE

Experimentation and the governing of socio-technical transitions

Harriet Bulkeley, Vanesa Castán Broto and Gareth A. S. Edwards

LONDON AND NEW YORK

First published 2015
by Routledge
2 Park Square, Milton Park, Abingdon, Oxon OX14 4RN

and by Routledge
711 Third Avenue, New York, NY 10017

Routledge is an imprint of the Taylor & Francis Group, an informa business

British Library Cataloguing in Publication Data
A catalogue record for this book is available from the British Library

Library of Congress Cataloging in Publication Data
Bulkeley, Harriet
An urban politics of climate change: experimentation and the governing of socio-technical transitions / Harriet A Bulkeley, Vanesa Castán Broto and Gareth A. S. Edwards.
pages cm
Includes bibliographical references and index.
1. Urban ecology (Sociology)—Government policy. 2. Climatic changes—Government policy. 3. Urban policy—Environmental aspects. I. Title.
HT241.B85 2014
307.76—dc23
2014025137

ISBN: 978-1-138-79109-1 (hbk)
ISBN: 978-1-138-79110-7 (pbk)
ISBN: 978-131-5-76304-0 (ebk)

Typeset in Bembo
by Book Now Ltd, London

CONTENTS

ILLUSTRATIONS

Figures

Tables

Box

ACKNOWLEDGEMENTS

The idea for this book began to germinate in 2007, when Harriet applied to the UK's Economic and Social Research Council for a Climate Change Fellowship. The support subsequently provided by the ESRC for the *Urban Transitions: climate change, global cities and the transformation of socio-technical networks* (Award Number: RES-066–27–0002) project has been critical in enabling the scale and scope of the work that underpins the book; supporting Vanesa and Gareth as Research Associates as well as the PhD studentships of Jon Silver and Gerald Taylor Aiken. The work of the project was supported directly by a number of associated PhD students and Research Associates – Andrea Armstrong, Cat Button, Sara Fuller, Andrés Luque-Ayala, and Anne Maassen – and we are grateful for their input and support. In addition, our work has benefited from the insights of a number of colleagues both near and far, including Michelle Betsill, JoAnn Carmin, Steve Graham, Simon Guy, Mike Hodson, Matthew Hoffmann, Kristine Kern, Simon Marvin, Adrian Smith and Heike Schroeder, as well as the conversations and perspectives generated by participants in the workshops organised by the Urban Transitions project in May 2009 in Manchester and March 2012 in Durham.

Producing a book is never a straightforward process. Our thanks also to the editorial team at Routledge for their support for this project, and particularly to Faye Leerink and Andrew Mould for their encouragement. We are particularly grateful to Jon Silver (Chapter 6), Andrés Luque-Ayala (Chapter 8) and Sara Fuller (Chapter 10) for their co-authorship of these chapters in the book, and to Andrea Armstrong for her dedicated editorial work on the draft manuscript. We would also like to thank Biodiversity Conservation India Ltd (BCIL) (India), DEAR Architects (Mexico), Sustainable Energy Africa, the Greater London Authority, Berlin Partner / virtualcitySYSTEMS GmbH, Sara Fuller, Jonathan Silver and Andrés Luque for their permission for the reproduction of photos and graphs in the book.

Book writing is not only a public endeavour, but also a very personal one. Over the course of the past seven years, our partners and children – three of whom have lived all of their relatively new lives since the start of our work together – have shown much forbearance in the face of the absences that academic work imposes on family life, both to places far away for fieldwork or conferences and to the kinds of distractions that new ideas and impending deadlines create. We thank them all for their patience and continued support, and hope that they can share our pride in the end result.

PART I

1

CLIMATE GOVERNANCE AND URBAN EXPERIMENTS

Enter the Anthropocene

Over the past two decades we have lived through a significant transformation in the way we think about ourselves and our relationship with the environment. For Earth Scientists, this is akin to a new geological period called 'the Anthropocene' in which 'the capability of contemporary human civilization to influence the environment at the scale of the Earth as a single, evolving planetary system' (Steffen *et al.* 2011: 842) is the dominant force. Climate change refers to the means through which anthropogenic emissions of greenhouse gases (GHGs) have come to alter atmospheric composition and in turn affect climatic conditions. Over the past twenty years scientists, policy-makers and the public have gradually moved away from asking 'is climate change happening?' to ask instead 'what can we do to stop the dangerous effects of climate change?' The multi-scalar nature of climate change, where GHGs are emitted locally but their impacts are transmitted through global processes, challenges the capacity of states to protect their citizens and the climate (Biermann and Dingwerth 2004; Bulkeley 2005; Schulz 2010). While scholarly and political attention has focused on how nation-states can seek to overcome this common action problem, there has been a growing awareness of the need to move 'beyond the state' in order to understand and develop the possibilities of a response to climate change.

The empirical evidence derived from diverse social and economic sectors has fostered debates about the type of action that should be prioritized, who should take responsibility for leading climate change action and where such action should be located (Hoffmann 2011). These responses are not confined to particular realms or levels of government, but rather take place in a multilevel governance context which extends vertically from international and transnational organizations to states, regions and cities and horizontally to civil society organizations, businesses and other non-state actors (Bulkeley and Betsill 2005; Romero Lankao 2012; Coutard and Rutherford 2010; Puppim de Oliveira 2009). From corporate climate change champions to community-based renewable energy schemes, interventions to decarbonize supermarket supply chains to revised planning policies, climate change has been put to work across a vast array of potential sites for political action.

Across the landscape of climate governance, increasing attention is being paid to the ways in which cities might respond to the challenges of mitigation, reducing emissions of GHGs, and adapting to the impacts of a changing climate (UN-Habitat 2011; World Bank 2010). The UN estimates that while the total world population will grow by 2.3 billion people between 2011 and 2050 (up to 9.3 billion), urban populations will grow even faster, increasing by 2.6 billion people up to 6.3 billion (UN Population Division 2011). Urban populations are both growing naturally and through processes of urbanization as people migrate from rural areas, seeking to embrace the economic opportunities cites provide. The consequences of this trend are geographically uneven. Rapid urban growth, particularly in cities in the global south, is often associated with the growth of settlements in high-risk areas, most vulnerable to global environmental change (e.g. Hardoy and Pandiela 2009; Parnell and Walawege 2011). Climate change-related events and challenges to resource security exacerbate a set of ongoing urban vulnerabilities resulting from growth, poor service provision, lack of access of vulnerable populations to basic services and weak governance structures (Watson 2009). If cities are vulnerable to the impacts of climate change, however, they are also at the heart of the processes through which GHG emissions are produced. While the exact proportion of GHG emissions that can be attributed to cities may be impossible to quantify (UN-Habitat 2011), the production and use of energy and transportation in cities, as well as the consumption of food and resources, means that cities are critical to any response that seeks to reduce GHG emissions. In one calculation, for example, the International Energy Agency (IEA) (IEA 2009: 21) finds that over two-thirds of annual global energy demand comes from cities and towns, which also 'produce over 70% of global energy-related carbon dioxide emissions'. By 2030, the IEA predicts that, with 60 per cent of the world's population, cities and towns will consume over 75 per cent of the world's energy demand, and that 'over 80% of the projected increase in demand above 2006 levels will come from cities in non-OECD countries' (IEA 2009: 21). Urban areas therefore present major challenges both in terms of concentrating activities that produce GHG emissions and concentrating vulnerabilities. At the same time, they also offer great potential to open up spaces for action addressing climate change. Over the past two decades, researchers and practitioners have sought to gather evidence across a range of cities to understand how and why cities are responding to climate change and the challenges they face.

In this book we seek to expand our knowledge and understanding of urban climate change responses by offering a new perspective on these issues. Theorizing urban climate governance as a set of processes that exceed the institutional boundaries of the local state, we argue that central to the urban response to climate change is a mode of *experimentation* where municipalities, private and civil society actors seek to demonstrate, experience, learn and challenge what it might mean to respond to climate change through a multiplicity of interventions, projects and schemes. Such experiments, we argue, are not simply *ad hoc* ventures, but need to be understood as situated and purposive interventions that demonstrate the ways in which new forms of authority are emerging in the context of climate change and the critical *socio-technical* dimension to realizing any governance response.

This book analyses the phenomenon of urban climate experimentation through the development of a new analytical approach and a series of cases that interrogate experimentation in practice. It seeks to develop our knowledge of the possibilities and limitations of urban responses by examining their emergence, dynamics and consequences. In the remainder of this chapter, we develop our argument that experimentation represents a critical aspect of the urban response to climate change. First, we situate our approach within the broader field of research on cities and climate change, exploring the roles of municipalities and non-state actors in responding to climate change within the urban arena as well as the broader context of urban political economy within which such responses have been pursued. We then turn to consider the ways in which different literatures have addressed questions of innovation and intervention in the urban arena. These insights inform an empirical analysis of evidence gathered from a survey of 100 global cities about the nature and characteristics of urban climate experiments. The chapter concludes by outlining the structure of the book.

Climate change: an emerging urban paradox?

Cities lie at the heart of the challenge of addressing climate change. Urban areas concentrate economic activities and social life which produce GHG emissions. While, as we point out above, several commentators have pointed out that anthropogenic GHG emissions are concentrated in urban areas, others point to the complexity of allocating emissions across geographical regions. With limited data and significant uncertainties, these studies have suggested that there is significant diversity within and between cities in terms of the production of GHG emissions, so any broad brush generalization will be inaccurate (Dodman 2009; Satterthwaite 2010; Hoornweg *et al.* 2011; Glaeser and Kahn 2010; Hillmer-Pegram *et al.* 2012). Furthermore, the urban contribution to climate change does not only rest with the in-situ production of GHG emissions, but also extends to their roles as centres of consumption for goods and services produced elsewhere. How, where and why cities should mitigate climate change is therefore no straightforward matter. At the same time, the concentration of people and hazards in urban areas poses significant challenges in terms of addressing the impacts of climate change (Dodman and Satterthwaite 2008). As with mitigation, exposure to climate risks differs across cities, with lower classes being generally more vulnerable to climate impacts (Grineski *et al.* 2012). Vulnerability to such impacts is linked not only to exposure, but also to poverty, segregation, lack of access to fundamental services and inequalities in terms of access to fundamental services (Ayers 2009; Hardoy and Pandiela 2009; Laukkonen *et al.* 2009). Responding to these challenges not only offers potential for addressing climate change, but also for attending to the wider causes and consequences of urban vulnerability. However, deciding what should be adapted, where, when and by whom raises fundamental questions in terms of the evidence base upon which decisions should be made, financial calculations, and questions of justice. Furthermore, while much discussion in urban planning for climate change has focused on whether adaptation or mitigation should be prioritized (e.g. Pizarro 2009), in practice both mitigation and adaptation share two crucial aspects that, at least within the city, suggest the need for convergence across both forms of intervention. First, both mitigation and adaptation require a definite intervention in the systems of provision and the built environment in the city (e.g. Coutard and Rutherford 2010; Mondstadt 2009; Laukkonen *et al.* 2009; Bulkeley *et al.* 2013), where GHG emissions and vulnerabilities to climate risks are profoundly interconnected. Second, both mitigation and adaptation are linked with broader questions of environmental justice and unequal access to resources and services in urban areas.

Responding to climate change at the urban level is not therefore simply a matter of recognizing that cities have an important role in terms of mitigation and adaptation and enacting a set of well-worn interventions. Rather, it raises critical issues of knowledge, politics and justice. These challenges and opportunities have generated considerable interest about the potential to foster and maintain effective action for addressing climate change through cities. Initially, much of the interest concentrated on understanding the motivations, reactions and latent institutional potential within local governments. Debates about multilevel governance have also generated interest in the ways in which non-state actors engage in climate change action. These debates tended to regard climate change as a separate sphere of urban governance. More recently, research has interrogated the ways in which climate change is becoming integral to the processes and politics of urbanism, through discourses such as ecological security, decarbonization and green growth. Below, we review each of these debates in turn.

Municipalities and climate change governance

Since the early 1990s, academic research on urban responses to climate change has explored the questions of why cities should take the leadership for climate change action, the forms of urban response that municipalities were undertaking, and the institutional factors which facilitate or prevent climate change action in city governments. These debates developed following the observation that municipal governments were taking action in the absence of initiatives at the national level and, in some cases such as the USA and Australia, in spite of it (Bulkeley and Betsill 2003; Gore and Robinson 2009; Warden 2011). This spurred interest in why municipal governments regarded cities as a key site to address climate change. This research points first to institutional reasons, second to reasons related to the potential to address climate change in cities, and third to reasons related to the adequacy of the responses at the city level. Those who invoke institutional reasons often also advocate the subsidiary principle, that is, the need to address policy problems at institutional levels as close as possible to citizens, so that interventions can lead to climate change-responsive planning that is also democratically legitimate (Kithiia and Dowling 2010). Commonly, institutional reasons are also used to identify the role that municipal governments have in the delivery of public services and management of infrastructure systems, so that they can intervene directly to reduce emissions and vulnerabilities (Bulkeley and Betsill 2003; Dodman and Satterthwaithe 2008). Moreover, emphasis on establishing measurable objectives at the local level helps establish direct linkages between policy aspirations and practical implementation challenges (Burch 2010; Bulkeley and Betsill 2005). Connecting climate change dynamics with their local impacts may also help local governments to mobilize different publics in response to climate change (Hunt and Watkiss 2011). At the urban level, citizens may find it easier to establish clear channels of communication with municipalities than with regional or state governments. Institutional motivations may also be embedded in the influence of other levels of government across vertical multilevel governance structures. For example, cities may simply be pressed by national governments to take action (Schreurs 2010).

A second – and related – set of arguments points to the potential of cities to concentrate people, knowledge, capital and resources in such a way as to provide the potential to implement climate change action. These arguments point to the critical role of municipal government intervention both through formal competencies and by enabling local action. Municipal governments have varied roles in the delivering of energy, land use planning, water, urban development, waste and transportation services – all of which are key sectors through which responses to climate change need to be addressed (Bulkeley and Betsill 2003). From an economic point of view, cities offer the opportunity to achieve economies of scale, which may allow interventions not available in other contexts. For example, in terms of achieving emissions reductions, cities may be characterized by high-energy density and service heterogeneity that may support the emergence of economies of scale and agglomeration, leading to social and technological innovation (Schulz 2010). Some local governments may also see climate change action as a way to realize additional economic and social benefits, for instance by promoting the so-called green economy, reducing air pollution, or creating new attractive urban areas, which may also provide a competitive advantage over other cities (Schreurs 2010). In this sense, initiatives such as the C40 Cities Climate Leadership Group, which promotes best practice action in key global cities, can be portrayed not only as a support network, but as building a competition space among cities promoting further action and innovation (Hodson and Marvin 2009). Third, in terms of the effectiveness of responses, commentators often highlight the potential of municipal government to tailor their responses to local needs and to use local knowledge to support decisions (Corfee-Morlot *et al.* 2011; Henstra 2012). For example, municipal governments have been identified as key actors in planning emergency responses to climate change because of their capacity to respond rapidly and coordinate the relationships between service providers, security services and communities (Crisp *et al.* 2012). Likewise, designing responses that include 'co-benefits' is thought to both improve effectiveness and add political legitimacy to the intervention (Betsill 2001; Koehn 2008).

From these three starting points, an increasingly broad range of scholarship has emerged that has sought to understand the sorts of responses that have taken place at the urban level. Historically, much emphasis focused on the development of plans to guide effective action. For example, since 1993, the pioneering Cities for Climate Protection (CCP) programme, led by the network of cities ICLEI (Local Governments for Sustainability), has reached hundreds of cities helping them to establish emissions inventories and to develop and implement local action plans accordingly. These initial plan-based approaches, most often found in medium-sized cities in North America, Europe and Oceania, were dominated by mitigation concerns (Bulkeley and Betsill 2003; Schreurs 2008; Zahran *et al.* 2008; Betsill and Bulkeley 2007). More recently, issues of urban resilience, vulnerability and adaptation have featured prominently in the policy agenda for a range of cities, from the concerted efforts of large cities, such as Durban and New York, to those who are members of the Asian Cities Climate Resilience Network and the UN-Habitat Cities and Climate Change Initiative. In this arena, debates have mostly been concerned with issues of risk assessment, the ways in which resilience and adaptation are – and are not – becoming mainstreamed within urban development, planning and disaster risk reduction, and the nature and extent of community involvement (Carmin *et al.* 2012; Solecki *et al.* 2012).

This work has shown the potential within municipalities to deliver action on climate change. Municipal responses have often based climate change action on previous experiences dealing with other environmental problems, such as environmental pollution, an emphasis on the need to achieve co-benefits and the implementation of approaches which pose low risks for municipalities (see for example Bulkeley and Betsill 2003; Bulkeley 2010). From an adaptation perspective, municipal interventions to implement broad structural changes improving building codes, urban design and infrastructure capacity provide direct opportunities to realize co-benefits such as improving life quality and public health (Bambrick *et al.* 2011). Well-known technologies such as combined heat and power or retrofitting with energy-efficient appliances have also been found to have potential to achieve emissions reductions at larger scales than anticipated without posing significant challenges to existing municipal policies (Schulz 2010; Castán Broto 2012). The strategies through which such measures are undertaken can be considered in terms of different 'modes of governance' (Bulkeley and Kern 2006; Kern and Alber 2008; Bulkeley *et al.* 2011). According to this analysis, municipal governments intervene through self-governing, regulation, provision and enabling actions. Self-governing includes action directed at regulating the operation of the government, from improving the energy efficiency of their own buildings to improving emergency procedures. These actions enable municipal governments to lead other actors by example. Regulations are directed at establishing rules or laws to govern conduct, from building codes to traffic conventions. Provision, in contrast, refers to interventions in the supply of public services, such as implementing renewable sources of energy or improving the efficiency of the water supply. Finally, enabling refers to support activities, ranging from financial incentives to the provision of knowledge, that facilitate the implementation of concrete actions by other actors in the city. The analysis reveals both a variety of forms of climate change intervention among municipal governments and the ways in which municipal governments attempt to enrol other urban actors in collective action on climate change.

In addition to seeking to understand how and why municipalities are responding to climate change, this body of work has also sought to examine the factors that shape the possibilities and limitations of these urban responses. Here, most of the focus has been on issues of institutional capacity and design. Carmin *et al.* (2012) capture this debate by demonstrating that municipal responses are influenced by both exogenous and endogenous factors that affect institutional capacity. In terms of exogenous factors, perhaps the most salient has been the influence of national governments and the implementation of regulations following international agreements (Schreurs 2010). City networks have also been effective in fostering and supporting climate change action at the urban level (Betsill and Bulkeley 2004; Kern and Bulkeley 2009). In addition, spectacles such as sporting events can trigger action by building public support and opening up the opportunity to draw resources and capital into the initiative (Bulkeley *et al.* 2009). Cities have been seen taking action to reduce emissions in anticipation of a world event, such as the World Cup in South African cities (which fostered carbon sequestration and greening programmes in Soweto) or the Beijing Olympics. Extreme weather events may also put pressure on municipal governments to implement adaptation actions (Romero Lankao 2012).

Among the contextual circumstances that facilitate or impede action, key endogenous factors are often related to the effectiveness of the municipal bureaucracy, available resources and political priorities (Bulkeley 2010). A first set of issues concerns the institutional capacity of local governments, in terms of being conferred an appropriate remit to support their actions in different sectors, having autonomy to act, and accessing appropriate skills to deal with the uncertainty and shifting nature of climate change demands (Romero Lankao 2012; Castello 2011). Municipal officials, for example, struggle to manage climate change information due to both lack of locally specific and relevant information (Burch 2010) and the difficulties inherent to dealing with heterogeneous resources within a limited pool of skills (Feliciano and Prosperi 2011; Betsill and Bulkeley 2007). The prevalence of technocratic models of action has also been identified as a barrier to effective climate action when these models are embedded in the institutional culture (Barbour and Deakin 2012). A second key set of factors underpinning the municipal response to climate change concerns resources, which are often absent or inadequate for the scale and complexity of the climate change problem (Satterthwaite 2008). In this context, the economic viability and cost-effectiveness of measures adopted – particularly with regards to their potential co-benefits – is a key aspect of implementing adaptation and mitigation measures. For example, policy makers interviewed about the retrofitting of heritage buildings in Hong Kong highlighted cost-effectiveness as the main limiting factor preventing them from moving from the visualization of a project to its implementation (Yung and Chan 2012).

Third, political will and leadership are regarded as central to effective urban climate change action (Bulkeley 2010; Romero Lankao 2012), and should be understood as the manifestation of political priorities that emerge within specific urban political contexts. For example, the comparison of sustainable transport policies in Stockholm (Sweden) and New Delhi (India) shows how urban political cultures shape radically different policies because:

> [w]hile in Stockholm, the extensive middle class, having satisfied its desires for car ownership, is gingerly stepping into worries about sustainability and global warming, in Delhi the demand of the motorists for removing congestion irrespective of efficient bus operations might halt the modernization of public transport.
>
> *(Thynell et al. 2010: 427)*

The support of locally-organized environmental groups is also mentioned as a key factor which may move a municipal government to commit to or implement climate change action (Puppim de Oliveira 2011; Zahran *et al.* 2008; Sharp *et al.* 2011). At the same time, political support can be garnered from beyond the municipality itself. While local governments can act relatively independently, their effectiveness depends on both the level of recognition afforded by the system (Gore 2010) and the capacity of local governments to extend their operation through city networks (Gustavsson *et al.* 2009). In this context, the importance of moving away from considering municipalities in isolation and contemplating instead the wide array of actors which intervene in the governing of climate change within the city has become central to the research agenda.

The role of non-state actors

The broad literature on the governance of climate change has grown increasingly interested in how to govern climate change 'beyond the state' (Biermann and Pattberg 2008; Okereke *et al.* 2009). Although municipalities have themselves been placed in this category, as part of the response to climate change that exceeds that of national governments, research has started to examine in more detail the multiplicity of actors involved in governing climate change in the city (Bulkeley and Schroeder 2012). While the literature on low carbon cities has generally emphasized the role of private actors – given their role in the provision of services and infrastructure – studies of adaptation have largely highlighted the role of communities in developing locally-specific adaptation strategies.

A key issue in this literature is the rationale for private sector involvement in urban responses to climate change. For example, a study in Hong Kong found that 'a small number of corporations are motivated to undertake climate action due to drivers such as competitive advantage, reputation, senior management leadership, risk reduction and external regulatory pressure' (Chu and Schroeder 2010: 301). Some may be attracted by existing government policies that create incentives to attract capital flows, such as in the case of the development and implementation of low carbon policies for cities in China (Colombier and Li 2012). Although the private sector is most often considered from the point of view of its role in mitigation initiatives, attracting private capital for climate and disaster risk reduction is gaining currency (Brugmann 2012). A study in Australia, for example, highlighted the growing interest among developers in incorporating adaptation concerns into their operation (Taylor *et al.* 2012). Overall, as with municipalities, the effectiveness of private actors in delivering GHG emission reductions or reducing climate vulnerability is thought to depend not only on endogenous factors but also their ability to draw in the support of other actors across the city and build legitimacy upon social embeddeness (Whiteman *et al.* 2011; see also Spath and Rohracher 2012).

Communities, including a variety of local stakeholders (residents associations, environmental pressure groups, community-based organizations) have been given credit not only for structuring interventions around locally-specific conditions but also because they often prioritize simple measures and lifestyle transformations over technological 'fixes' and technocratic approaches which do not target the underlying physical and social changes required (Knuth 2010). Initiatives such as Transition Towns, whose underlying motivation is challenging society's dependence on fossil fuels, emphasize an holistic approach which engages with the overall sustainability of the community and the lifestyles of its members. This represents a shift from technological to social and political innovation. In the context of adaptation, '[o]nly organized communities can reflect and act on members' priorities, learn from these experiences and build more effective and targeted interventions' and thus, 'Community-Based Adaptation' (CBA) is portrayed as 'an effective mechanism for reducing vulnerability among low-income and marginalized groups' (Dodman *et al.* 2010: 20). CBA is particularly important when public institutions lack capacity to respond effectively to extreme events (Wamsler and Lawson 2012), and may help to direct attention to existing low-cost strategies that communities may have developed in the context of risk (Jabeen *et al.* 2010; Porio 2011). However, there is also recognition that local government support greatly enhances community-based interventions for climate change adaptation (Ahammad 2011).

Beyond the focus on the roles of individual actors, this research has also highlighted the emergence of boundary organizations, or intermediary actors, which broker relationships between previously disconnected actors and realms of potential response (Hodson and Marvin 2012). Boundary organizations may, for example, emerge within government to support higher levels of government accessing localized forms of decision-making, for instance through local authorities or deliberative practices (Corfee-Morlot *et al.* 2011). Universities are increasingly seen as such boundary organizations, brokering the knowledge needs from municipal governments and other actors (Hillmer-Pegram *et al.* 2012). Overall, there is emphasis on the capacity of leading agents, often municipalities, to draw together a broad network of actors who can collectively address complex climate change challenges by generating multiple ideas, negotiating interests and establishing a degree of momentum required to overcome inertia (e.g. Hardoy and Pandiela 2009; Henstra 2012; Rutland and Aylett 2008). Work in this field suggests that there is considerable heterogeneity both in terms of the actors that are seeking to govern climate change in the city, and also in terms of the diversity of responses and roles that they adopt in these processes. This in turn raises questions about the extent to which such forms of governance may be symptomatic of wider shifts within urban political economies.

Climate governance as urban governance

Urban climate governance exceeds the boundaries of the municipality. Over the past two decades, municipal actors have engaged in forms of network politics (Bulkeley 2005) that extend their reach and influence through collaboration in transnational networks and that confer 'a significant role in the diffusion of both techniques and norms, important functions of governance and politics' (Toly 2008: 345). Networks, such as C40, ICLEI Cities for Climate Protection, the ACCRN and UN-Habitat's CCCI, have mobilized municipal actors in collaboration with the private sector, not-for-profit organizations, communities and the global development funding community, creating a mesh-like set of structures through which to govern climate change at the urban level. The emergence of these new forms of transnational authority has taken place alongside a growing realization of the limitations of the multilateral process of inter-state negotiation and co-operation in delivering climate change responses (Hoffmann 2011), and broader political economic processes that are sometimes described in terms of the 'fragmentation' of political authority and the neoliberalization of the economy. The urban governance of climate change is both reflective of these trends, and a means through which they are being realized. Climate change responses at the urban level are therefore critical to the reconfiguration of urban governance. Hodson and Marvin (2009), for example, have coined the term 'urban ecological security' to highlight how strategic attempts to safeguard resource flows 'are now informing strategies to reconfigure cities and their infrastructures in ways that help to secure their ecological and material reproduction' (2009: 193). In the context of the emergence of new urban logics of carbon control, research has drawn attention to 'the ways in which urban development politics and the strategies of urban regimes might be influenced not simply by broader "economic and regulatory relations" but also by new environmental agendas such as climate change' (Jonas *et al.* 2011: 2538; see also While *et al.* 2010). Such approaches point to the ways in which climate change is being mobilized at once as an economic development strategy and as a means through which a new kind of urban politics is constituted. The production of urban climate change responses requires not only the definition of the problem space and the constitution of political and economic interests, but also distinct forms of intervention. Rice (2010) uses her analysis of climate governance in Seattle to demonstrate how this process involves the dual production of urban territories on the one hand and the materialities of climate change on the other:

> … a rearticulation of territorialization with respect to climate change is occurring in Seattle, one that is not concerned with defining a territorially based 'inside' and 'outside' but with the attribution of material nature to specific places and activities that are within the boundaries of state institutions.
>
> *(Rice 2010: 938)*

Urban climate governance responds to a global agenda while signalling a wider shift in the nature of political authority and in the political economy of carbon. New rationales of resource security, carbon control and securing cities from risk are being generated alongside and through urban climate change responses. This suggests that, rather than being a matter of discrete policies and plans, individual measures or projects, climate change has come to matter at the urban level in a more fundamental way than just addressing its causes and impacts.

The urban politics of innovation

Understanding the ways in which climate change has come to matter at the city level requires not only engaging with specific issues of adaptation and mitigation, but with the broader structural dynamics of capitalist globalization and competitive advantage that shape the nexus between climate change and urbanization (Whitehead 2013). Understanding this connection requires engaging with cities as 'bundles of political and economic processes that transcend metropolitan space' (Whitehead 2013: 1353). Within urban studies more broadly, such approaches have documented the ways in which cities are critical sites for both domination and contestation, and in which there is little space amongst urban elites for alternative views of everyday urban lives and what a city does or ought to do. For Roy and Ong (2011) the political economy of globalization has served to exclude urban variation and particularity, despite the fact that 'caught in the vectors of particular histories, national aspirations, and flows of cultures, cities have always been the principal sites for launching world-conjuring projects' (Roy and Ong 2011: 1). Through examples in Hong Kong, Dubai and Delhi, they argue that this logic of globalization has now come to embody the 'worlding' of cities (Roy and Ong 2011). This notion describes the means through which the urban milieu is made global through particular forms of intervention and, as they put it, experiments that simultaneously reinvent both urban space and what counts as being global. In their drive to physically remake the city and reinvent the particularities of certain cities in global discourses, these worlding practices define the city as a field of intervention where contemporary problems are addressed in multifarious ways (Roy and Ong 2011). Such interventions 'promiscuously draw upon ideas and objects, and find allies in multiple sources that are recontextualized for resolving urban problems' (Roy and Ong 2011: 4). Characterized as an 'unprecedented phenomenon' (Hebbert and Jankovic 2013), climate change is one problem that cities are tackling through worlding practices of the kind discussed above. Yet, in engaging with climate change, cities are not just caught in the dynamics of intervention and problem definition: they also mediate innovation. Change itself, mediated through the production of new knowledges and understandings of the city, becomes an objective of urban intervention.

Understanding the practices and politics of intervention in the city can therefore help analyse how the global agenda develops by and through the urban, and how the urban at the same time is transformed into a global project. Historically, the literature on cities has drawn on their recognition as sites for innovation and knowledge production. More recently – and reflecting the concerns above for the ways in which the politics and economies of cities transcend particular places – researchers have sought to examine the ways in which knowledge and policy travel across different urban sites and shape the nature of urban interventions. In a parallel debate, scholars have examined the nature of innovation within infrastructure systems as a means of achieving systemic change. This work draws attention to the socio-technical nature of infrastructure systems, and to the ways in which innovation emerges within, and has the potential to reconfigure, such networks. It draws attention to the ways in which forms of urban innovation and intervention are connected to broader political, economic and material processes. Rather than being a technical matter, both bodies of work also point to the ways in which innovation is assembled through both social and material means mediated by multiple forms of agency. In the next section, we discuss the implications of these debates for our own analysis, focusing first on the potential of socio-technical transition debates to contribute to our understanding of urban responses to climate change, and second on the relevance of the policy mobilities literature for understanding the governance of climate change in the city.

Socio-technical transitions and the city

The literature on systems innovation and sustainability has sought to understand how transitions take place in the regimes that structure the workings of large technical systems or infrastructures. Socio-technical regimes are regarded as stable configurations of social and technical components, which rely on a set of formal rules and informal conventions to align the interests of different actors and existing technological possibilities (see Elzen *et al.* 2004; Geels 2005, 2007; Grin *et al.* 2010). Two important assumptions of this large body of theory can be traced back to Hughes' (1983) work on the electrification of western society. First, innovations are thought to be arranged in systems which include 'interacting components of different kinds, such as the technical and the institutional, as well as different values' (Hughes 1983: 6). Second, challenging traditional engineering understandings of technical systems, these socio-technical systems are 'neither centrally controlled nor directed towards a clearly defined goal' (Hughes 1983: 6). Instead, an array of actors intervene within a regime, the loose configuration of materials and actors bounded by institutions and rules, in which two patterns of change are identified: a slow path in which different elements of the regime change gradually until its overall renewal; or an abrupt, rapid change – a transition – in which the disturbance of socio-technical relationships within the regime cause a rapid realignment. In relation to climate change and sustainability, Smith *et al.* (2010) argue that the allure of these perspectives resides in broadening both the problem framing – by engaging with both the technological and institutional aspects of transitions – and the analytical framing, moving beyond coevolutionary economics. However, their emphasis on agency and paradigmatic innovation (Coenen *et al.* 2012) explains the success of systems innovation analyses in driving much sustainability policy, as one that relies on constructing and giving recipes for the possibility of change.

Within systems innovation, the multilevel perspective provides insights into how systems innovation drives transitions (see Elzen *et al.* 2004; Geels 2005; Geels and Schot 2007). It proposes an analytical model of socio-technical systems divided in levels with increased degrees of stability. In this model, the socio-technical regime describes the relatively stable configuration of technical and social elements which is re-arranged during the transition. Regimes occur within a more structured landscape, over which actors have less agency. In contrast, niches emerge outside the mainstream regimes, where innovations resulting from actors' agency can flourish, posing a challenge to the incumbent regime. The model explains the obduracy of socio-technical systems and the potential for innovation to break through relatively stable systems. In this analysis, niches emerge as key spaces were innovation can be deliberately fostered by key actors in the regime, so that they can bring about a systemic, rapid and abrupt transformation towards low carbon energy systems.

As case studies, cities have been prominent in the development of systems innovation studies. Hughes (1983), for example, understood cities such as London, Chicago and Berlin as centres for socio-technical change, embedded in histories of institutional development and utility buildings that shape both technological histories and the lives of Londoners, Chicagoans and Berliners. However, the systems innovations literature and, in particular, the multilevel perspective have been broadly criticized for overlooking the geographical dimensions of transitions. Coenen *et al.* (2012) argue that the neglect of the spatial dimensions of transitions in favour of its historical treatment reduces the comparability between contextually situated places, and tends to homogenize technological innovation systems to the extent that innovation is left at the mercy of naively constituted 'global' forces. Thus, they argue, the lack of territoriality may lead to the misleading suggestion that a 'sustainability transition may take place *anywhere*' (Coenen *et al.* 2012: 976, emphasis in the original). For Hodson and Marvin (2010) the contextual politics that unfold at the urban scale can be related to multilevel approaches to transitions theory through the question of governance and, in particular, the emergence and development of particular visions of socio-technical systems, which relate their changes with the actions of actors and institutions operating at the urban scale. They point out that engaging with the urban dimensions of transitions will require understanding the extent to which processes of territorial governance map onto the governance of the regime, in a bid to delineate territorial and regime properties. In line with this argument, Spath and Rohracher (2012) argue that the governance of transitions (and, in particular, the mobilization of groups of actors) is a crucial prerequisite for innovations to attain institutional embeddedness, and thus foster transitions. Further, they argue that regional and municipal policies provide linkages between niches and regimes, even when they do not explicitly target niche innovation (Spath and Rohracher 2012). The role of urban and regional authorities in providing exemplars (Hodson and Marvin 2009) or forms of paradigmatic development (Coenen *et al.* 2012; Spath and Rohracher 2012) suggests that deliberate attempts to bring about transitions within an urban context should be regarded as part of the palette of worlding practices that emerge from the problematization and reinvention of the city. However, systems innovation scholarship is yet to develop a spatial perspective on urban innovation and transitions which recognizes their historically- and spatially-situated character. While such perspectives bring a critical dimension to understanding urban climate change responses as interventions in larger socio-technical systems, of the importance of innovation in processes of transition, and of the ways in which this is conducted socially and materially, they lack a sufficient engagement with the urban as a distinct political, economic and social realm.

Urban innovation and policy mobilities

Scholarship on policy mobilities proposes a different approach to examine, comparatively, the ways in which urban interventions come to be made and understood, with a particular focus on the circulation of knowledge as giving rise to distinct practices of innovation in cities. Closely related to the emerging thought on worlding cities, the literature on policy mobilities is broadly interested in 'understanding how and with what consequences urbanism is assembled through policy actors' purposive gathering and fixing of globally mobile resources, ideas, and knowledge' (McCann and Ward 2012: 43). This follows on from an understanding of the city as a place of intervention (Roy and Ong 2011). Here, policy is a shared problem, which is framed and responded to within a specific arena (Freeman 2012). The literature on policy mobilities thus looks into how policies travel from context to context, embedded in discourses and practices, and by 'following the policy' provides the basis for a territorial and context-based analysis of the development of prototypes and innovations (Peck and Theodore 2012).

These ideas have clear relevance for understanding the governance of climate change in cities, which has clearly been largely shaped by attempts to transpose policy from one context to another, using model cities as examples and establishing baselines that may enable the comparison between cities with networks such as ICLEI or C40 (Kern and Bulkeley 2009). The literature on policy mobilities places such processes under critical scrutiny, questioning the extent to which such transpositions can be understood through the medium of rational policy making and emphasizing the messy actualities of mobilizing policy in response to global issues such as climate change (Cochrane and Ward 2012). Yet, in understanding global policy circulations, this body of work also points out how any specific intervention (or policy) unfolds in a given milieu, is determined not just by location but also by the operations of multiple knowledge and institutional agents. Specifically, a diverse group of experts and middling bureaucrats that, although operating in a specific urban context, learn and develop calculative practices through interactions in global, regional or local networks (Cochrane and Ward 2012; see also Roy 2012). Larner and Laurie's (2010) example of the privatization of telecoms and water, for example, shows that travelling engineers draw from the global circulation of discourses while also shaping these circulations through their own movement. In looking at how discourses in circulation take root in urban space, the literature on policy mobilities connects with Li's analysis of how policy programmes come to be assembled and put to work in the case of rural development in Indonesia (Li 2007), a debate to which we return in Chapter 2.

While the literature on socio-technical transitions highlights the importance of recognizing the role of localized forms of agency in reassembling urban technological systems, the literature on policy mobilities highlights how such agency is constituted both locally and globally. Following this, we argue that it is critical to engage with how climate change narratives travelling from global to local realms and vice versa can shape and are shaping the city looking at purposeful interventions in specific contexts while simultaneously rejecting any notion that such interventions are determined locally. At the same time, the focus of the socio-technical transitions literature on the assemblage of interventions through the conjunction of particular materialities, techniques, social networks and narratives provides a starting point for engaging with the processes through which such interventions come to be constituted and realized. While policies and knowledges may travel, this work suggests that it is as important to consider how particular technologies and techniques are mobilized, and how the materialities and infrastructures of particular contexts enable and constrain the possibilities and pathways of intervention. Each approach adds weight to the importance of considering the strategic interventions in the governing of the city. Our emphasis is on the intention to intervene and, in particular, on deliberate strategic attempts to shape the city in particular ways in response to climate change. In doing so, our work develops a spatial conceptualization of urban transitions, on the one hand, and engages with the city as a site of purposive intervention, on the other. Yet, a focus on urban reconfiguration requires not only mapping the actors who bring about change – and how they do it – but also an understanding of how this change is materialized in specific urban patterns and life practices, an issue to which we turn in the next section

An urban world of climate experimentation

The efforts of municipalities and non-state actors to address climate change have created a plethora of climate change plans, policies, initiatives and projects across an array of diverse urban contexts. While some have come to regard this proliferation of fragmented responses as symptomatic of a *lack* of urban governance, in this book we take a different starting point. Rather than assuming that the fragments of a climate response visible within cities are either simply curiosities, of little real value, or that they need to be stitched together in order to form a coherent response, we suggest that *they are integral to the governing of climate change in cities* (Bulkeley and Castán Broto 2013). We argue that such interventions can be regarded as *climate change experiments*, where experiment means the 'action of trying anything, putting it to proof', as well as 'to have experience of ... to feel' (OED Online 2013). Climate change experiments provide a means through which what responding to climate change entails and for whom can be tried out, put to the test and experienced. Drawing these responses into the frame of analysis enables us to interrogate *how* the urban governing of climate change is taking place and to investigate its dynamics and consequences.

The notion of experimentation as a form of response to environmental problems has been explored in various literatures (Bulkeley and Castán Broto 2013). As detailed above, innovation and experimentation has been central to explanations of how and why urban policies travel and become embedded. Such understanding, albeit recast in relation to the dynamics of globalization and the political economies of worlding cities, draws on a longer heritage of interest in policy experimentation within public policy and political science. Of particular relevance, Hoffman (2011) examines the emergence of governance experimentation in which multiple public and private actors adopt innovative forms and means for addressing climate change. Experimentation, Hoffman (2011) argues, is a response to both the moribund process of multilateral negotiation and the broader trend of the fragmentation of authority and driven by the opportunities for developing new sources of legitimacy, finance and moral authority by a range of actors. Despite its relevance to the climate change issue, Hoffman's concept of experimentation, like others within this broad tradition, is confined to the policy realm and does not engage with the ways in which experimentation takes on a material quality through the reconfiguration of particular sites and infrastructure systems, leaving it inadequate for interrogating how governing takes place socially and materially.

Innovation – and, in particular, niche forms of experimentation – have also been central to the debates on how and why infrastructure networks may undergo transition. Experiments and niche innovations create protected spaces where new technological or social innovations can gain momentum, and may eventually break through to disrupt the regime. However, as we discussed above, while this literature demonstrates how the intersection of material and social factors is central to the process of intervention, there has been limited consideration of the ways in which socio-technical systems are constituted at the urban level. A third body of work has explicitly engaged with the notion of cities as urban laboratories, showing that strategic attempts have been made to experiment in specific urban contexts (Evans 2011; see also Evans and Karvonnen 2011). Yet, the notion of the laboratory focuses the analysis on the creation of spaces of containment and exception in the city and their dynamics, which may obscure the way in which such interventions are structured in relation to global environmental politics and political economy (Bulkeley 2005).

Given these limitations, we find that none of these approaches can, by themselves, adequately account for the phenomenon of urban climate change experimentation. Instead, we seek to develop a new framework for analysis situated at the confluence of these three theoretical traditions, as purposive, strategic attempts to configure urban socio-technical systems, which constitute a critical form of urban governance. Under this framework, urban climate change experiments are defined according to three criteria: first, an intervention is experimental when it is purposive and strategic but recognizes the open-ended nature of socio-technical processes; second, an intervention is a climate change experiment if its purpose is to reduce emissions of greenhouse gases (mitigation) and/or vulnerabilities to climate change impacts (adaptation); third, a climate change experiment is urban when it is delivered by or in the name of an existing or imagined urban community. To understand the extent and diversity of climate change experimentation (as defined above) both in the global North and the global South, we undertook a comparative analysis focused on 100 global cities (Bulkeley and Castán Broto 2013; Castán Broto and Bulkeley 2013). Using secondary sources, we compiled a database of 627 experiments found in these cities in order to investigate the nature and characteristics of experiments in different sectors. Below, we summarize the findings from this analysis, before exploring the implications for understanding how and why experimentation is emerging as a form of urban climate governance.

Urban climate change experimentation trends: a comparison

Analysis of our database shows that experimentation is a relatively recent trend: almost 80 per cent of the experiments recorded started after the ratification of the Kyoto Protocol in 2005. This suggests that urban climate change experimentation can be characterized as part of the growing interest in climate change during the past decade and the growing sense of urgency that has translated the concerns from international negotiations to local development and sustainability agendas. Experiments are likely to emerge in a variety of sectors but they are particularly common in urban infrastructure (31 per cent), built environment (25 per cent), and transport (19 per cent; see Figure 1.1). Most of the experiments in the database were mitigation experiments, with only 12 per cent focused on adaptation. This may be related to the historically lower levels of engagement with adaptation at the international level and in urban transnational networks, or because it is often regarded as an issue to be considered in most operations and not always differentiated from on-going disaster management programmes (Satterthwaite 2008). As a result, adaptation initiatives are not necessarily undertaken purposively in the name of climate change and may therefore be considered climate change experiments according to our definition upon which the database was built.

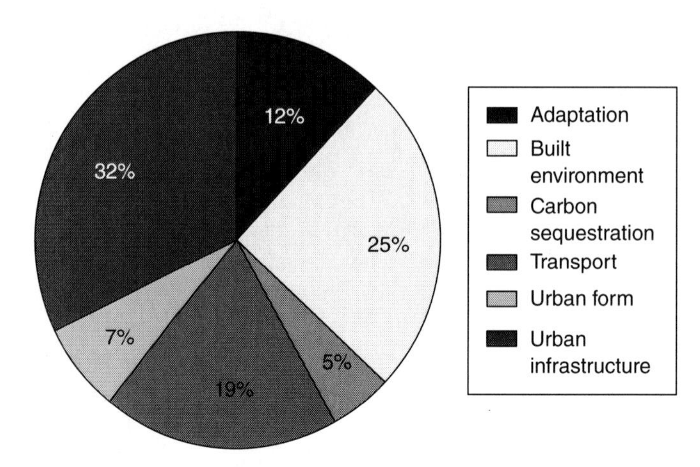

FIGURE 1.1 The distribution of experiments by sector.

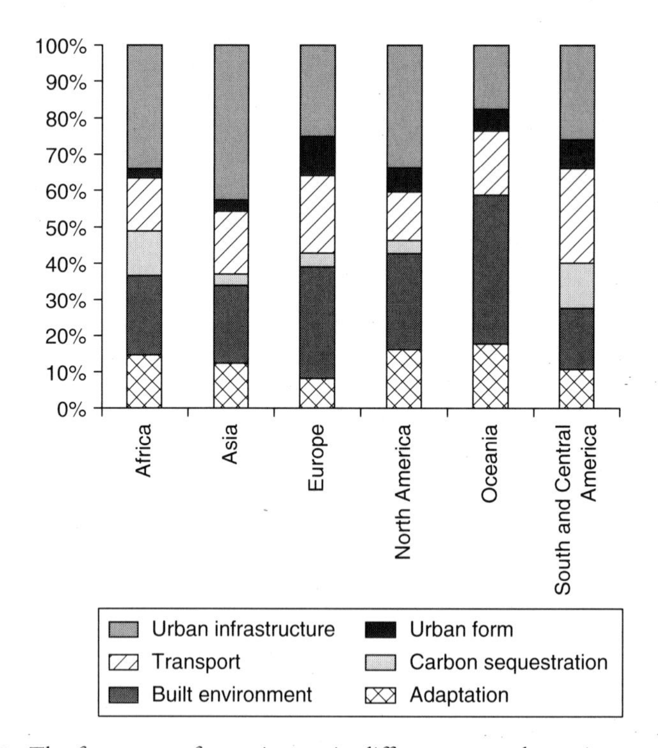

FIGURE 1.2 The frequency of experiments in different sectors by region.

In terms of where experimentation is taking place, perhaps surprisingly and in contrast to the existing literature, analysis demonstrates that experiments are not confined to particular regions of the world in either geographically or economic development terms (Figure 1.2). Experiments appear equally likely in cities in developed countries, but also in cities in developing regions, which had previously been thought of as lacking the resources or the motivation to undertake climate change action, especially mitigation. Through statistical regression analysis, we evaluated whether the emergence of experiments depended on any or multiple city characteristics (city growth, total population, density, area, GDP, role in the World Economy and participation in city networks) (Castán Broto and Bulkeley 2012). Only this last factor – participation in city networks – appeared to explain a significant percentage of the variation in the emergence of experiments in a particular city. These results point to the difficulties in understanding the causes and patterns of urban climate change experimentation. Belonging to city networks is an explanatory factor, supported by the qualitative evidence from examples of experiments directly associated with the commitments made to these networks, for example, in London through its participation in the C40 network (Hodson and Marvin 2007). However, the direction of causation is unclear – do cities that contain experiments join transnational networks, or does involvement in transnational networks lead either to greater levels of experimentation or more readily accessible records of such experimentation taking place? Furthermore, this evidence suggests that urban climate change experimentation goes beyond international policy initiatives, or the size and concentration of resources or population in any one city. Understanding where and why experiments emerge requires looking beyond trends into the specific context where such experiments emerge, a task which we begin in the remaining chapters of this book.

When it comes to understanding who is governing climate change at the urban level, in line with previously-gathered evidence through case-study research, the analysis shows that local governments have a prominent role, leading 66 per cent of urban climate change experiments (Figure 1.3). However, looking at experiments also reveals that alongside city governments other actors may be playing a key role in climate change governance. Private actors lead 15 per cent and civil society actors lead 9 per cent of initiatives. The database also suggests that partnership is a key feature of climate change governance: 296 experiments (47 per cent) involved some form of formally-recognized partnership between actors at different

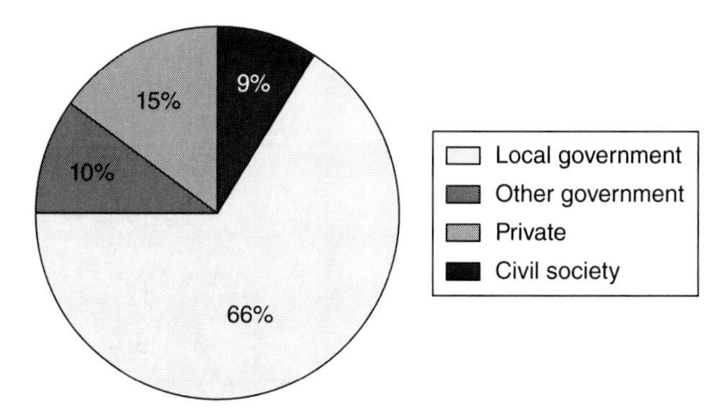

FIGURE 1.3 The proportion of climate change experiments led by actors in different sectors of society.

governance levels, both in terms of vertical governance (e.g. partnerships between local, regional and national governments) and horizontal (e.g. partnerships between governments, civil society organizations and private actors). Partnerships are important for local governments because they can be used, for example, as a means to facilitate further action for climate change taken by other actors and extend the operation of the state (Kern and Bulkeley 2009). Beyond the local government, partnerships are generally considered a key tool for capacity building (Eakin 2006) and building consensus (Newman *et al.* 2009). Looking across the types of partnerships, we also observe that they offer the opportunity for participation to a diverse array of actors. The dominance of partnerships resonates on the one hand with the growing importance of enabling modes of governance in which governing is conducted not only by local authorities but also by other actors who seek to facilitate experiments led by other actors and the growing influence of boundary and intermediary organizations that establish linkages across different actors.

Considering the ways in which experimentation takes place, the database demonstrates its often explicitly socio-technical nature. Urban climate change experiments are not limited to either technological or policy innovations; rather, they purposively attempt to change both the material arrangements and the cultures, norms and conventions that determine collective GHGs emissions and climate-related vulnerabilities in the city. In the database, technical forms of innovation are more prevalent (present in 76 per cent of all experiments) than social ones (present in 50 per cent). The emphasis on technological innovation, especially in sectors such as urban infrastructure, points to the ways in which governing climate change has an important material dimension. Moreover, the results of the database confirm previous observations that energy efficiency issues dominate local climate change agendas (Bulkeley *et al.* 2009; see also Bulkeley and Kern 2006; Rutland and Aylett 2008). Up to 45 per cent of the database experiments seek to intervene in energy systems, and include efforts aimed at reducing demand as well as new systems of energy production and generation in the urban infrastructure sector. As well as these material concerns, there is also evidence that climate change experimentation is linked to issues of wider politics. About a quarter of experiments reported a concern with issues either of the distribution of harm or benefit from climate change, or with social inclusion issues, but, surprisingly, local authorities were less likely to express these ideals, perhaps because other private and civil society actors may draw on these arguments are a means to legitimize their intervention in the climate change agenda.

Experiments and urban transitions in context

Our analysis demonstrates the growing importance of urban climate change experiments. The growing empirical significance of this phenomenon is such that we can no longer treat them as isolated cases of good practice, but must instead attend to what they mean for governing climate change in the city as well as to their potential consequences for urbanism. Yet the analysis above is limited in the degree to which it enables us to understand the significance of these experiments within their urban context, both in terms of explaining how they intervene in the reconfiguration of urban socio-technical systems and the extent to which they are able to catalyse broader systemic transitions in response to climate change.

Attending to these gaps in our knowledge is vital if we are to understand the potential – and limitations – of experimentation as a mode for achieving a rapid change or transition in response to climate change. On the one hand, there is evidence that experimentation can lead to broad change. For example the Young Cities research project (2005–2013) that built a 35-ha energy-efficient and resilient urban settlement in the southern part of Hashtgerd, Iran, became a real-life pilot project that has proven the efficacy of specific actions and eventually was transferred into a legally binding comprehensive plan (Seelig 2011). The localized development of innovation can lead to '(i) levering technological exploration; (ii) providing room for experimentation and testing and (iii) creating ground for exploitation and demonstration of new technologies' (Carvalho *et al.* 2012: 388). Equally, the potential of experiments may lie in their capacity to act in the interplay between global governance networks and locally implemented demonstrations, so that they may help develop an understanding of how to bring about and legitimize alternative socio-technical configurations (Spath and Rohracher 2012).

Yet, the potential for transformation is constrained by the extent to which such change leads to more progressive forms of urbanism or instead 'the language of transitions is mobilized by narrow coalitions of interest that do not result in genuine, radical transition but, rather, work to reproduce the economic status quo' (Hodson and Marvin 2012: 437). Moreover, interventions leading to urban transitions for climate change have led to substantial redistributive effects, suggesting that impacts on inequality are a key aspect to consider in climate change policy (Gusdorf *et al.* 2008). Given that, from a socio-technical perspective, any form of transition or 'change must be understood as operating both through and within heterogeneous spatial contexts as well as over many varied time scales' (Coutard and Rutherford 2010: 723), understanding the potential, politics and consequences of experimentation requires an approach that can move beyond broad trends and undertake 'more in-depth research exploring and analysing the spatially and temporally differentiated processes and practices of transition within specific contexts' (Coutard and Rutherford 2010: 723). Drawing from this emerging body of work, in the rest of this book we develop an approach for the research and analysis of urban climate change experiments which emphasizes: their strategic and political nature; their material character and their capacity to create new socio-technical configurations; and their potential and limitations in advancing urban futures that are environmentally and socially just. In so doing, we argue that such experiments may lead to socio-technical transformation in response to climate change not because of their scope or scale, or even through their transfer from one place to another, but because they provide a means of reordering and reconfiguring the systems and structures through which urbanism is constituted.

Chapter overview

In the rest of this book, we develop our argument for the importance of understanding experimentation as central to the urban governing of climate change. In *Chapter 2*, we establish the conceptual foundations to understanding governing as a (re)configuration of socio-technical and political-ecological relations in the city. From this starting point, we argue that climate change experiments are forms of intervention that enable us to explore the ways in which climate change is becoming a 'problem' in cities that requires intervention and improvement (Li 2007), and to consider how the rationalities, techniques and practices of experimentation both bring forth new possibilities for acting on climate change, but, because of their emergent characteristic, are always open to contestation and failure. Understanding experimentation as part of the governing of climate change in this manner, the chapter then turns to the dynamics of experimentation itself, introducing a threefold analytic framework through which to understand this mode of intervention in terms of the making, maintaining, and living of experiments. The chapter concludes by considering the implications of this framework for how we understand the unfolding politics and practices of governing climate change in the city.

Part II of the book puts this framework to work through the detailed analysis of eight cases of climate experiments in different urban contexts. *Chapter 3* examines the case of the Towards Zero Carbon housing development on the outskirts of Bangalore, India, an experiment targeted at realizing low carbon housing for India's emerging middle class, detailing its emergence, the ways in which it has been maintained both as a site of experimentation and enrolled into the political-ecological circuits of the city, and the extent to which it has been able to foster new forms of 'low carbon' identity in the everyday practices of residents and practitioners across the city. *Chapter 4* focuses on experimentation with low-carbon housing for a low-income neighbourhood in Monterrey, Mexico in the project ViDA (Vivienda de diseño ambiental). The chapter analyses the failure to maintain the experiment in-situ while all the while extending its relation with broader circuits of housing finance across Mexico in the shape of the Green Mortgage, and examines how and why the lived realities of the experiment have diverged from its original intentions. In *Chapter 5*, we turn to a case of retrofitting housing and energy conservation in the US city of Philadelphia: the 'Retrofit Philly "Coolest Block" Contest'. In bringing together public and private interests to create a competition for a retrofit or weatherization prize, this case allows us to examine the ways in which the nature and purpose of housing is redefined in the light of climate change, and to consider the consequent implications for how such interventions may serve to challenge or sustain existing forms of urban inequality. *Chapter 6* turns to a second case of housing retrofit, the provision of insulated ceiling installations in privately-owned housing in Mamre, Cape Town, through a partnership between Danish International Development Assistance (DANIDA) and led by the City of Cape Town. The chapter documents the making of the experiment, and considers the extent to which it can be maintained in the context of limited finances and capacity constraints, and its implications in terms of realizing the broader objectives of social justice and development.

In *Chapter 7*, we examine the ways in which experimentation is taking place in the energy system of one city, Berlin, Germany, through the Solar Atlas and the Solardachbörse (Solar Roof Exchange) initiatives. We find that, in the context of distributed capacity, strained finances, an energy system focused on heating provision and a housing system dominated by landlords, it has been difficult to embed solar energy within the political and material circulations of the city. While the experiment has been successfully assembled, despite the powerful private interests it has enrolled, it has yet to be maintained within these broader forms of circulation, and remains distant from the everyday practices of citizens. *Chapter 8* focuses on the case of Sao Paulo to examine solar hot water (SHW) experiments in social housing that are creating a new model for energy provision urban infrastructure. The chapter considers how solar hot water experiments have been assembled and sustained within the social housing sector by the Housing and Urban Development Company of the State of São Paulo (CDHU) – Brazil's largest housing agency – and examines the extent to which they have become normalized within everyday practice and extended across the urban landscape. *Chapter 9* examines the emergence of parallel experiments designed to act on individual behaviour and energy use in the city of Hong Kong, China: PowerSave, promoted through Friends of the Earth; and Climateers, led by WWF. We find that experimentation functions in this case both to deflect attention from state-based responsibilities for action, but also as a means of advancing a politics of mitigation where the state is absent; in treading this fine line, experimentation may serve to promote new forms of climate justice, but also to sustain inequalities. In *Chapter 10*, we analyse the development of the 'low carbon zone' in Brixton, London, as a means for intervening in the forms of energy provision and consumption in the city. Analysing the practice of the low carbon zone, and its relation to the development and emergence of Transition Town Brixton – an alternative site of energy politics in the borough – the chapter reflects on the implications of such experiments for the politics of low carbon transitions.

In *Part III* we draw together the findings and conclusions from the empirical cases to consider the consequences of climate change experiments and the implications for our understanding of urban climate change governance. In *Chapter 11* we turn our attention to how experimentation is caught up in the politics of justice in urban climate change governance. We argue that an urban politics of justice is essential in addition to the existing international perspectives on climate justice, and then examine how notions of justice are manifested in climate change experiments. Experiments are emergent and unruly, subject to conflicting intentions and configured through the socio-material and ecological dynamics of particular urban conditions as they unfold. While we find in some cases that powerful social interests are able to channel experiments towards particular ends, this is not exclusively a neoliberalizing project seeking to create private good at the expense of social benefit. Furthermore, experimentation remains under contestation, through everyday forms of resistance and engagement as well as from the interests of elites concerned to marginalize climate concerns in the city. There is no straightforward urban climate politics aligned for or against vested interests, but rather we find that experimentation contains the possibilities for both progressive and regressive forms of urban transition. In *Chapter 12*, the concluding chapter, we consider the implications of the cases for our understanding of the politics and practices of urban climate governance. We revisit the conceptual arguments advanced at the beginning of the book to assess the extent to which experiments provide a means through which governing is conducted, and to assess the socio-material and political-ecological means through which this is taking place across the multiple cases. In so doing, we problematize the argument that experiments need to either be 'scaled up' or 'transferred' in order to have significant implications, but consider instead the multiple ways in which they are mobilized to achieve particular social, economic and political ends. We argue that, rather than representing a wholesale shift from one system state to another, we can expect climate change transitions to be fragmented, contested and prone to failure as well as success. At the same time, we argue that the presence and embedding of climate change experiments in the city is not something that can be undone; climate change is now an urban issue.

References

Ahammad, R. (2011) 'Constraints of pro-poor climate change adaptation in Chittagong City'. *Environment and Urbanization*, 23(2): 503–15.

Ayers, J. (2009) 'International funding to support urban adaptation to climate change'. *Environment and Urbanization*, 21(1): 225–40.

Bambrick, H. J., Capon, A. G., Barnett, G. B., Beaty, R. and Burton, A. (2011) 'Climate change and health in the urban environment: adaptation opportunities in Australian cities'. *Asia-Pacific Journal of Public Health*, 23: 67S–79S.

Barbour, E. and Deakin, E. A. (2012) 'Smart Growth Planning for Climate Protection Evaluating California's Senate Bill 375'. *Journal of the American Planning Association*, 78(1): 70–86.

Betsill, M. M. (2001) 'Mitigating climate change in US cities: opportunities and obstacles'. *Local Environment: The International Journal of Justice and Sustainability*, 6(4): 393–406.

Betsill, M. M. and Bulkeley, H. (2004) 'Transnational networks and global environmental governance: the Cities for Climate Protection program'. *International Studies Quarterly*, 48: 471–93.

——(2007) 'Looking back and thinking ahead: a decade of cities and climate change research'. *Local Environment*, 12(5): 447–56.

Biermann, F. and Dingworth, K. (2004) 'Global environmental change and the nation state'. *Global Environmental Politics*, 4(1): 1–22.

Biermann, F. and Pattberg, P. (2008) 'Global environmental governance: taking stock, moving forward'. *Annual Review of Environment and Resources*, 33(1): 277–94.

Brugmann, J. (2012) 'Financing the resilient city'. *Environment and Urbanization*, 24(1): 215–32.

Bulkeley, H. (2005) 'Reconfiguring environmental governance: towards a politics of scales and networks'. *Political Geography*, 24(8): 875–902.

——(2010) 'Cities and the governing of climate change'. *Annual Review of Environment and Resources*, 35: 229–53.

Bulkeley, H. and Betsill, M. M. (2003) *Cities and climate change: urban sustainability and global environmental governance*. London: Routledge.

——(2005) 'Rethinking sustainable cities: multilevel governance and the "urban" politics of climate change'. *Environmental Politics*, 14(1): 42–63.

Bulkeley H. and Kern K. (2006) 'Local government and climate change governance in the UK and Germany'. *Urban Studies*, 43: 2237–59.

Bulkeley, H. and Castán Broto, V. (2012) 'Urban experiments and climate change: securing zero carbon development in Bangalore'. *Contemporary Social Science*: 1–21.

——(2013) 'Government by experiment? Global cities and the governing of climate change'. *Transactions of the Institute of British Geographers*, 38: 361–75.

Bulkeley, H. and Schroeder, H. (2012) 'Beyond state/non-state divides: global cities and the governing of climate change'. *European Journal of International Relations*, 18(4): 743–66.

Bulkeley, H., Schroeder, H., Janda, K., Zhao, J., Armstrong, A., Chu, S.Y. and Ghosh, S. (2009) *Cities and climate change: the role of institutions, governance and urban planning*. Report for the World Bank Urban Research Symposium: Cities and Climate Change.

Bulkeley, H., Schroeder, H., Janda, K., Zhao, J., Armstrong, A., Chu, S. and Ghosh, S. (2011) 'The role of institutions, governance and planning for mitigation and adaption by cities'. In *Cities and climate change: responding to an urgent agenda*. New York: The World Bank, pp. 68–88.

Bulkeley, H., Castan Broto, V. and Maasen, A. (2013) 'Low-carbon transitions and the reconfiguration of urban infrastructure'. *Urban Studies* (published online 27 August 2013).

Burch, S. (2010) 'Transforming barriers into enablers of action on climate change: insights from three municipal case studies in British Columbia, Canada'. *Global Environmental Change*, 20(2): 287–97.

Carmin, J., Anguelovski, I. and Roberts, D. (2012) 'Urban climate adaptation in the Global South: planning in an emerging policy domain'. *Journal of Planning Education and Research*, 32(1): 18–32.

Carvalho, L., Mingardo, G. and van Haaren, J. (2012) 'Green urban transport policies and cleantech innovations: evidence from Curitiba, Goteborg and Hamburg'. *European Planning Studies*, 20(3): 375–96.

Castán Broto, V. (2012). 'Social housing and low carbon transitions in Ljubljana, Slovenia'. *Environmental Innovation and Societal Transitions*, 2: 82–97.

Castán Broto, V. and Bulkeley, H. (2013) 'A survey of climate change experiments in 100 cities'. *Global Environmental Change*, 23(1): 92–102

Castello, M. G. (2011) 'Brazilian policies on climate change: the missing link to cities'. *Cities*, 28(6): 498–504.

Chu, S. Y. and Schroeder, H. (2010) 'Private governance of climate change in Hong Kong: an analysis of drivers and barriers to corporate action'. *Asian Studies Review*, 34(3): 287–308.

Cochrane, A. and Ward, K. (2012) 'Researching the geographies of policy mobility: confronting the methodological challenges'. *Environment and Planning A*, 44(1): 5–12.

Coenen, L., Benneworth, P. and Truffer, B. (2012) 'Toward a spatial perspective on sustainability transitions'. *Research Policy*, 41(6): 968–79.

Colombier, M. and Li, J. (2012) 'Shaping climate policy in the housing sector in northern Chinese cities'. *Climate Policy*, 12(4): 453–73.

Corfee-Morlot, J., Cochran, I., Hallegate, S. and Teasdale, P. J. (2011) 'Multilevel risk governance and urban adaptation policy'. *Climatic Change*, 104: 169–97.

Coutard, O. and Rutherford, J. (2010) 'The rise of post-network cities in Europe? Recombining infrastructural, ecological and urban transformation in low carbon transitions'.

In Bulkeley, H., Castán Broto, V., Hodson, M. and Marvin, S. (eds) *Cities and low carbon transition*. Abingdon: Routledge, pp. 107–25.

Crisp, J., Morris, T. and Refstie, H. (2012) 'Displacement in urban areas: new challenges, new partnerships'. *Disasters*, 36: S23–42.

Dodman, D. (2009) 'Blaming cities for climate change? An analysis of urban greenhouse gas emissions inventories'. *Environment and Urbanization*, 21(1): 185–201.

Dodman, D. and Satterthwaite, D. (2008) 'Institutional capacity, climate change adaptation and the urban poor'. *IDS Bulletin-Institute of Development Studies*, 39(4): 67–74.

Dodman, D., Mitlin, D. and Co, J. R. (2010) 'Victims to victors, disasters to opportunities: community-driven responses to climate change in the Philippines'. *International Development Planning Review*, 32(1): 1–26.

Eakin, H. (2006) *Weathering risk in rural Mexico: climatic, economic and institutional change*. Tucson, AZ: University of Arizona Press.

Elzen, B., Geels, F. W. and Green, K. (2004) *System innovation and the transition to sustainability: theory, evidence and policy*. Cheltenham: Edward Elgar.

Evans, J. (2011) 'Resilience, ecology and adaptation in the experimental city'. *Transactions of the Institute of British Geographers*, 36: 223–37.

Evans, J. and Karvonnen, A. (2011) 'Living laboratories for sustainability: exploring the politics and epistemology of urban transition'. In Bulkeley, H., Castan Broto, V., Hodson, M. and Marvin, S. (eds) *Cities and low carbon transitions*. London: Routledge, pp. 126–41.

Feliciano, M. and Prosperi, D. C. (2011) 'Planning for low carbon cities: reflection on the case of Broward County, Florida, USA'. *Cities*, 28(6): 505–16.

Freeman, R. (2012) 'Reverb: policy making in wave form'. *Environment and Planning A*, 44(1): 13–20.

Geels, F. (2005) *Technological transitions and system innovations, a co-evolutionary and socio-technical analysis*. London: Edward Elgar.

——(2007) 'Analyzing the breakthrough of rock 'n roll (1930–1970): multi regime interaction and reconfiguration in the multi-level perspective'. *Technological Forecasting and Social Change*, 74: 1411–31.

Geels, F. and Schot, J. (2007) 'Typology of sociotechnical transition pathways'. *Research Policy*, 36(3): 399–417.

Glaeser, E. L. and Kahn, M. E. (2010) 'The greenness of cities: carbon dioxide emissions and urban development'. *Journal of Urban Economics*, 67(3): 404–18.

Gore, C. (2010) 'The limits and opportunities of networks: municipalities and Canadian climate change policy'. *Review of Policy Research*, 27(1): 27–46.

Gore, C. and Robinson, P. (2009) 'Local government response to climate change: our last, best hope?' In Selin, H. and Van Deveer, S. D. (eds) *Changing climates in North American politics: institutions, policymaking and multilevel governance*. Cambridge, MA: MIT Press, pp. 138–58.

Grin, J., Rotmans, J. and Schot, J. (2010) *Transitions to sustainable development: new directions in the study of long term transformative change*. London: Routledge.

Grineski, S. E., Collins, T. W., Ford, P., Fitzgerald, R., Aldouri, R., Velazquez-Angulo, G., Romo Aguilar, M. and Lu, D. (2012) 'Climate change and environmental injustice in a bi-national context'. *Applied Geography*, 33(1): 25–35.

Gusdorf, F., Hallegatte, S. and Lahellec, A. (2008) 'Time and space matter: how urban transitions create inequality'. *Global Environmental Change-Human and Policy Dimensions*, 18(4): 708–19.

Gustavsson, E., Elander, I. and Lundmark, M. (2009) 'Multilevel governance, networking cities, and the geography of climate-change mitigation: two Swedish examples'. *Environment and Planning C: Government and Policy*, 27: 59–74.

Hardoy, J. and Pandiella, G. (2009) 'Urban poverty and vulnerability to climate change in Latin America'. *Environment and Urbanization*, 21: 203–24.

Hebbert, H. and Jankovic, V. (2013) 'Cities and climate change: the precedents and why they matter'. *Urban Studies*, 50(7): 1332–47.

Henstra, D. (2012) 'Toward the climate-resilient city: extreme weather and urban climate adaptation policies in two Canadian provinces'. *Journal of Comparative Policy Analysis*, 14(2): 175–94.

Hillmer-Pegram, K. C., Howe, P. D., Greenberg, H. and Yarnal, B. (2012) 'A geographic approach to facilitating local climate governance: from emissions inventories to mitigation planning'. *Applied Geography*, 34: 76–85.

Hodson, M. and Marvin, S. (2007) 'Understanding the role of the national exemplar in constructing "strategic glurbanization"'. *International Journal of Urban and Regional Research*, 31(2): 303–25.

——(2009) '"Urban ecological security": a new urban paradigm?'. *International Journal of Urban and Regional Research*, 33(1): 193–215.

——(2010) *World cities and climate change: producing urban ecological security*, Milton Keynes: Open University Press.

——(2012) 'Mediating low-carbon urban transitions? Forms of organization, knowledge and action'. *European Planning Studies*, 20(3): 421–39.

Hoffmann, M. J. (2011) *Climate governance at the crossroads: experimenting with a global response after Kyoto*. Oxford: Oxford University Press.

Hoornweg, D., Sugar, L. and Gomez, C. L. T. (2011) 'Cities and greenhouse gas emissions: moving forward'. *Environment and Urbanization*, 23 (1): 207–27.

Hughes, T. (1983) *Networks of power electrification in Western society, 1880–1930*. Baltimore, MD: Johns Hopkins University Press.

Hunt, A. and Watkiss, P. (2011) 'Climate change impacts and adaptation in cities: a review of the literature'. *Climatic Change*, 104: 13–49.

IEA (International Energy Agency) (2009) *Cities, towns and renewable energy: yes in my front yard*. Paris: IEA.

Jabeen, H., Johnson, C. and Allen, A. (2010) 'Built-in resilience: learning from grassroots coping strategies for climate variability'. *Environment and Urbanization*, 22(2): 415–31.

Jonas, A. E. G., Gibbs, D. and While, A. (2011) 'The New Urban Politics as a politics of carbon control'. *Urban Studies*, 48(12): 2537–54.

Kern, K. and Alber, G. (2008) *Governing climate change in cities: modes of urban climate governance in multi-level systems*. Proceedings of the OECD conference on Competitive Cities and Climate Change. Paris: OECD.

Kern, K. and Bulkeley, H. (2009) 'Cities, Europeanization and multi-level governance: governing climate change through transnational municipal networks'. *Journal of Common Mark. Studies*, 47: 309–32.

Kithiia, J. and Dowling, R. (2010) 'An integrated city-level planning process to address the impacts of climate change in Kenya: the case of Mombasa'. *Cities*, 27(6): 466–75.

Knuth, S. E. (2010) 'Addressing place in climate change mitigation: reducing emissions in a suburban landscape'. *Applied Geography*, 30(4): 518–31.

Koehn, P. H. (2008) 'Underneath Kyoto: emerging subnational government initiatives and incipient issue-bundling opportunities in China and the United States'. *Global Environmental Politics*, 8(1): 53–77.

Larner, W. and Laurie, N. (2010) 'Travelling technocrats, embodied knowledges: globalizing privatization in telecoms and water'. *Geoforum*, 41(2): 218–26.

Laukkonen, J., Blanco, P. K., Lenhart, J., Keiner, M., Cavric, B. and Kinuthia-Njenga, C. (2009) 'Combining climate change adaptation and mitigation measures at the local level'. *Habitat International*, 33: 287–92.

Li, T. M. (2007) *The will to improve: governmentality, development, and the practice of politics.* Durham, NC: Duke University Press.

McCann, E. and Ward, K. (2012) 'Assembling urbanism: following policies and "studying through" the sites and situations of policy making'. *Environment and Planning A*, 44(1): 42–51.

Monstadt, J. (2009) 'Conceptualizing the political ecology of urban infrastructures: insights from technology and urban studies'. *Environment and Planning A*, 41(8): 1924–42.

Newman, P., Beatley, T. and Boyer, H. (2009) *Resilient cities: responding to peak oil and climate change.* Washington, DC: Island Press.

OED Online (2013) *Oxford English Dictionary*. Online at http://www.oed.com/ (accessed 10 January 2014).

Okereke, C., Bulkeley, H. and Schroeder, H. (2009) 'Conceptualizing climate governance beyond the international regime'. *Global Environmental Politics*, 9(1): 58–78.

Parnell, S. and Walawege, R. (2011) 'Sub-Saharan African urbanization and global environmental change'. *Global Environmental Change – Human and Policy Dimensions*, 21: S12–S20.

Peck, J. and Theodore, N. (2012) 'Follow the policy: a distended case approach'. *Environment and Planning A*, 44(1): 21–30.

Pizarro, R. E. (2009) 'The mitigation/adaptation conundrum in planning for climate change and human settlements: introduction'. *Habitat International*, 33(3): 227–29.

Porio, E. (2011) 'Vulnerability, adaptation, and resilience to floods and climate change-related risks among marginal, riverine communities in Metro Manila'. *Asian Journal of Social Science*, 39(4): 425–45.

Puppim de Oliveira, J. A. (2009) 'The implementation of climate change related policies at the subnational level: an analysis of three countries'. *Habitat International*, 33(3): 253–59.

——(2011) 'Why an air pollution achiever lags on climate policy? The case of local policy implementation in Mie, Japan'. *Environment and Planning A*, 43(8): 1894–909.

Rice, J. L. (2010) 'Climate, carbon, and territory: greenhouse gas mitigation in Seattle, Washington'. *Annals of the Association of American Geographers*, 100(4): 929–37.

Romero Lankao, P. (2012) 'Governing carbon and climate in the cities: an overview of policy and planning challenges and options'. *European Planning Studies*, 20(1): 7–26.

Roy, A. (2012) 'Ethnographic circulations: space–time relations in the worlds of poverty management'. *Environment and Planning A*, 44(1): 31–41.

Roy, A. and Ong, A. (eds) (2011) *Worlding cities: Asian experiments and the art of being global.* Chichester, West Sussex, UK: Blackwell.

Rutland, T., and Aylett, A. (2008) 'The work of policy: actor networks, governmentality, and local action on climate change in Portland, Oregon'. *Environment and Planning D: Society and Space*, 26(4): 627–46.

Satterthwaite D. (2008) 'Cities' contribution to global warming: notes on the allocation of greenhouse gas emissions'. *Environment and Urbanization*, 20(2): 539–49.

——(2010) 'The contribution of cities to global warming and their potential contributions to solutions'. *Environment and Urbanization Asia*, 1(1): 1–12.

Schreurs, M. A. (2008) 'From the bottom up: local and subnational climate change politics'. *Journal of Environment and Development*, 17(4): 343–55.

——(2010) 'Multi-level governance and global climate change in East Asia'. *Asian Economic Policy Review*, 5(1): 88–105.

Schulz, N. (2010) 'Lessons from the London Climate Change Strategy: focusing on combined heat and power and distributed generation'. *Journal of Urban Technology*, 17(3): 3–23.

Seelig, S. (2011) 'A master plan for low carbon and resilient housing: the 35 ha area in Hashtgerd New Town, Iran'. *Cities*, 28: 545–56.

Sharp, E. B., Daley, D. M. and Lynch, M. S. (2011) 'Understanding local adoption and implementation of climate change mitigation policy'. *Urban Affairs Review*, 47(3): 433–57.

Solecki, W., Rosenzweig, C., Hammer, S. and Mehrotra, S. (2012) 'Urbanization of climate change: responding to a new global challenge'. In Sclar, E., Volavka-Close, N. and Brown, P. (eds) *The urban transformation: health, shelter and climate change*. London: Routledge, pp. 197–220.

Spath, P. and Rohracher, H. (2012) 'Energy regions: the transformative power of regional discourses on socio-technical futures'. *Research Policy*, 39(4): 449–58.

Smith, A., Voß, J-P. and Grin, J. (2010) 'Innovation studies and sustainability transitions: the allure of the multi-level perspective and its challenges'. *Research Policy*, 39: 435–48.

Steffen, W., Grinevald, J., Crutzen, P. and McNeill, J. (2011) 'The Anthropocene: conceptual and historical perspectives'. *Philosophical Transactions of the Royal Society A – Mathematical Physical and Engineering Sciences*, 369(1938): 842–67.

Taylor, B. M., Harman, B. P., Heyenga, S. and McAllister, R. R. J. (2012) 'Property developers and urban adaptation: conceptual and empirical perspectives on governance'. *Urban Policy and Research*, 30(1): 5–24.

Thynell, M., Mohan, D. and Tiwari, G. (2010) 'Sustainable transport and the modernization of urban transport in Delhi and Stockholm'. *Cities*, 27(6): 421–29.

Toly, N. J. (2008) 'Transnational municipal networks in climate politics: from global governance to global politics'. *Globalizations*, 5(3): 341–56.

UN-Habitat (2011) *Global report on human settlements: cities and climate change*. Nairobi, Kenya: UN-Habitat.

UN Population Division (2011) *World urbanization prospects*. Available online at http://www.un.org/en/development/desa/population/ (accessed December 2013).

Wamsler, C. and Lawson, N. (2012) 'Complementing institutional with localized strategies for climate change adaptation: a South-North comparison'. *Disasters*, 36(1): 28–53.

Warden, T. (2011) 'Viral governance and mixed motivations: how and why US cities engaged on the climate change issue, 2005–2007'. In Hoornweg, D., Freire, M., Lee, M. J., BhadaTata, P. and Yuen, B. (eds) *Cities and climate change: responding to an urgent agenda*. Washington DC: The World Bank, pp. 161–74.

Watson, V. (2009) 'The planned city sweeps the poor away …: urban planning and 21st century urbanization'. *Progress in Planning*, 72: 151–93.

While, A., Jonas, A. E. G. and Gibbs, D. (2010) 'From sustainable development to carbon control: eco-state restructuring and the politics of urban and regional development'. *Transactions of the Institute of British Geographers*, 35: 76–93.

Whitehead, M. (2013) 'Neoliberal urban environmentalism and the adaptive city: towards a critical urban theory and climate change'. *Urban Studies*, 50(7): 1348–67.

Whiteman, G., de Vos, D. R., Chapin, F. S., Yli-Pelkonen, V., Niemela, J. and Forbes, B. C. (2011) 'Business strategies and the transition to low-carbon cities'. *Business Strategy and the Environment*, 20(4): 251–65.

World Bank (2010) *Cities and climate change: an urgent agenda*. Washington, DC: World Bank.

Yung, E. H. K. and Chan, E. H. W. (2012) 'Implementation challenges to the adaptive reuse of heritage buildings: towards the goals of sustainable, low carbon cities'. *Habitat International*, 36(3): 352–61.

Zahran S., Brody, S. D., Vedlitz, A., Grover, H. and Miller, C. (2008) 'Vulnerability and capacity: explaining local commitment to climate change policy'. *Environment and Planning C: Government and Policy*, 26(3): 544–62.

2

GOVERNING THROUGH EXPERIMENTATION

Analysing the urban politics of climate change

Introduction

It has become increasingly common to regard climate change as an urban problem. Organizations ranging from development agencies to multinational corporations and community groups to municipal governments see both the threat of climate change and the promise of a response to it in cities. As Chapter 1 outlined, this growing engagement with climate change at the urban level by a multiplicity of actors has attracted the attention of both scholars and policymakers. While much of this attention has focused on understanding and designing urban climate change plans and strategies, these are far from being the only – or even dominant – urban responses to climate change. Multiple forms of urban intervention are now invested with the intention of addressing climate change, including low carbon houses, community-based adaptation schemes, cycleways, and behaviour change campaigns to name just a few. Analyses of urban climate change governance have tended to either side-line such interventions as isolated examples, demonstration projects or niche practices, or to consider them merely as the measures through which plans are implemented, all the while keeping their analytical eye trained on the process of policy development itself. In this Chapter, we advance an alternative conceptual approach through which to engage with the governing of climate change. This enables us to attend to not only *why* the governing of climate change occurs in a particular mode, but also *how* it is accomplished and the gaps and contradictions that emerge between the intention to govern and the messy actualities of governing in practice (Li 2007a).

In the first part of the Chapter, we survey the literature on the global governance of climate change and explore how urban responses have been situated conceptually within this approach. Framed predominantly in terms of actors, interests and institutions, this literature has analysed the city's role in the governing of climate change and the role of municipalities as actors within a multilevel, global system. In the light of some of the shortcomings of such approaches, we turn to critical political and social theory, and in particular the work of scholars of governmentality, to find more fruitful terrain for opening up the notion of governance and making space for an explicitly socio-material understanding of the city. Using this literature, we argue that the governing of climate change is orchestrated by the competing rationalities, techniques and tactics that actors mobilize as they seek to define both the object to be governed and the subjects with whom it is concerned. Adopting such a perspective means asking new questions about both the programmes and practices of governing, whilst also recognizing the varied sites and spheres through which governing is conducted – from strategic alignments to mundane forms of accounting, from dazzling low carbon developments to the ordinary electricity cable. As Chapter 1 demonstrated, we find that the constellation of interventions which together form the urban climate change governance landscape are often explicitly *experimental* in character, seeking to develop, trial and attest to the experience of responding to climate change. We suggest that such forms of experimentation can be regarded as part of the 'art of government' (Foucault 2009: 92–3) involved in the response to climate change.

Having established our approach to theorizing climate change governance, in the second part of the chapter we develop an analytical framework through which to apply these concepts to particular instances of urban climate change experimentation. We argue that experimentation involves three constituent elements, which we term *making*, *maintaining* and *living*. The *making* of experiments involves defining the problem and rendering it amenable to intervention through a process of assembling the discrete elements which facilitate action (for instance institutions, people, raw materials and capital). The *maintaining* of experiments involves deploying tactics and techniques to both establish the experiment in the face of the contradictions emergent between the intention to govern and its practice, and also to further embed it within the circulations of which it is a part. The *living* of experiments involves bringing into being distinct subjectivities which are charged with specific forms of conduct and come to be contested or simply fall into neglect. In closing the chapter, we reflect on the possibilities and implications of our approach for analysing the urban response to climate change.

Governing climate change

In the two decades since climate change emerged on the international agenda, debates concerning the ways in which it can and should be governed have ranged far and wide (Bulkeley and Newell 2010). Initially conceived within the discipline of international relations as a matter of common pool resources and inter-state negotiations, the field of global governance has sought to expand the range of actors, institutions and interests involved in the governing of climate change. The emergence of municipalities as one of a number of actors directly engaged in the governing of climate change beyond the nation-state has been central to the development of this body of work, problematizing the notion that climate change is an issue confined to the politics of international affairs. However, we suggest that the predominant focus on municipal responses to climate change through the lens of global governance has limited the analysis of why and how the governing of climate change takes place at the urban level. In this section, we first review the notion of global governance before advancing an alternative account for understanding the processes and practices of governing climate change which draws on critical social and political theory, particularly the bodies of work concerned with governmentality and urban political ecology (UPE).

Climate governance as global governance

In its broadest sense, governance 'relates to any form of creating or maintaining political order and providing common goods for a given political community on whatever level' (Risse 2004: 298) or 'to the modes and practices of the mobilization and organization of collective action' (Coafee and Healy 2003: 1979). Since the 1990s, there has been a visible 'governance turn' within the social sciences, reflecting the changing nature of political and social order in which what was regarded as state authority has been increasingly reconfigured between supra-national, sub-national and private bodies. A proliferation of approaches to, and means for, understanding governance have arisen as a result. In the global environmental domain, Biermann and Pattberg (2008) identify three distinct ways the term has been used: analytical, normative and critical. In the climate change arena, governance is most commonly used in its analytical guise – to identify and explain the ways in which climate change is governed. They suggest that three related phenomena are central to this analytical approach to global environmental governance:

> first, the emergence of new types of agency and of actors in addition to national governments, the traditional core actors in international environmental politics; second, the emergence of new mechanisms and institutions of global environmental governance that go beyond traditional forms of state led, treaty-based regimes; and third, increasing segmentation and fragmentation of the overall governance system across levels and functional spheres.
>
> *(Biermann and Pattberg 2008: 280)*

In some accounts, such new actors, mechanisms and spheres remain closely aligned to the development of the international process of climate politics, whilst in others 'the emergence of autonomous spheres of authority' beyond the state (Dingwerth and Pattberg 2006: 197) is central to the conceptualization of climate politics as a matter of global governance (Bulkeley and Newell 2010; Hoffmann 2011; Okereke *et al.* 2009). From this second perspective, the growing diversity and complexity of climate governance is not only a factor of the array of interests that have emerged around national negotiating positions and the development of international institutions, but also proceeds from a suite of responses to climate change that represent alternative forms of governance. As a result, global climate governance includes 'all purposeful mechanisms and measures aimed at steering social systems towards preventing, mitigating, or adapting to the risks posed by climate change' (Jagers and Stripple 2003: 385).

Together with the analysis of the growing private response to climate change amongst corporations and third sector organizations, examining the ways in which cities and regions are not only developing climate action individually but have also constituted transnational governance networks has been central to the development of the field of climate governance (see Chapter 1). The global governance literature has made space for considering cities both as part of the multilevel system of climate governance and as a valid sphere for action in their own right (Betsill and Bulkeley 2006; Burch 2010; Chu and Schroeder 2010; Schreurs 2010). For the most part, this work has focused on the ways in which municipalities – as actors and as institutions – have sought to govern climate change (Chapter 1). While such analyses have been critical in opening up the debate as to what constitutes climate governance, they have also been limited. Perhaps most fundamentally, the discussion of global climate governance has been preoccupied with debates 'on the changing roles and power of state and nonstate actors, and on the resulting changes in the institutionalization of political authority' (Sending and Neumann 2006: 653). This focus on what and who governs has to some extent been to the neglect of questions of why, how and for whom this is being undertaken (Bulkeley and Newell 2010). As a result, most accounts of climate governance pay little attention to 'the actual practices through which governance takes place' (Sending and Neumann 2006: 654) or to questions of who gains and who loses from such practices.

This is particularly relevant in the urban context, where inequality, both in terms of access to resources and agency to intervene in urban processes, is spatially concentrated and therefore highly visible. Yet studies of urban responses to climate change have tended to neglect these dynamics. Okereke *et al.* (2009) suggest that what is required to overcome these limitations is a critical engagement with some of the fundamental terms in which the debate on climate governance is cast, in particular: authority; power; structure-agency; and the specific practices through which governing is accomplished. Various scholars have turned to work within traditions of critical social and political theory, including political economy, actor-network theory and governmentality, in order to reconsider what governing climate change might entail (e.g. Blok 2012; Bulkeley and Newell 2010; Lövbrand and Stripple 2012; Oels 2005; Stripple and Bulkeley 2014). Our perspective in this book develops the notion of governmentality as a means to link strategic interventions, their attempts at controlling material contexts and populations and their consequences in terms of the distribution of authority. The next section explains how this perspective may influence the analysis of urban responses to climate change.

Urban climate governance as governmentality?

The understanding that to govern involves the formation and exercise of rule under conditions of authority in which those who are governed consent to this order underpins much of the literature on global governance (Dean 2007: 36). Historically, however, governing has also been used in a much broader sense to refer to 'any practice that more or less deliberately seeks to direct, guide or control others' or oneself (Dean 2007: 36). For Foucault, this form of governing (which he termed government) operates through 'the conduct of conduct'; that is, 'modes of action, more or less considered and calculated, that were destined to act upon the possibilities of action of other people' (Foucault 2000a: 341). Foucault locates the emergence of government as a mode of power during the development of the modern state in Europe alongside a concern to govern the population at large. Rather than being concerned with securing territory, the intent of government is to 'improve the condition of the population, to increase its wealth, longevity and its health' (Foucault 2009: 105). As Li (2007a: 5–6) observes, this 'will to improve the condition of the population, is expansive' and thus tends to encompass multiple spheres of social life. Accomplishing government as the improvement of the population therefore 'requires the exercise of what Foucault identified as a distinct, governmental rationality' (Li 2007a: 6) and a suite of accompanying techniques and arts of government that are referred to by the broad notion of governmentality.

Foucault (2009: 108) defines governmentality as 'the ensemble formed by institutions, procedures, analyses and reflections, calculations and tactics that allow the exercise of this very specific, albeit very complex, power'. Governmental rationalities provide the means through which the object of government – what Lockwood and Davidson (2010: 391) refer to as its 'ends, means and limits' – are established and pursued (Foucault 2000a: 344). Equally critical are 'the techniques, the practices, that give concrete form to this new political rationality' (Foucault 2000b: 410), which as Murdoch (2000: 505) observes 'both make rationalities "visible" and permit their extension through time and space'. Such practices are 'not just governed by institutions, prescribed by ideologies' but also 'up to a point, possess their own specific regularities, logic, strategy, self-evidence and "reason"' (Foucault 2000c: 225). Governmentality operates through the 'right way of arranging (disposer) things in order to lead (conduire) … not to the form of the "common good" … but to a "suitable end", an end suitable for each of the things to be governed' (Foucault 2009: 99).

Equally significant is Foucault's insistence that governing is not concerned only with the social world of actors and institutions, but rather with 'a sort of complex of men and things' (Foucault 2009: 96). Such things, Foucault goes on, include 'wealth, resources, means of subsistence, and of course, the territory with its borders, qualities, climate, dryness, fertility … customs, habits, ways of acting and thinking … accidents, misfortunes, famine, epidemics and so on' (Foucault 2009: 96). Seen in this way, governing becomes a socio-material practice, one that operates through the assemblage of social, technical and natural elements in any one domain. Thus, each configuration of power requires an established technological regime and both technologies and the expertise that informs them are required for the maintenance and reproduction of power. In this sense, each technological conquest represents a strategy-led struggle between diverse forces to rearrange a particular assemblage (cf. Swyngedouw 2004).

Governmentality provides a particular perspective on power and insights into the ways in which it is exercised. When the processes and practices of urban responses to climate change are approached from this perspective, it allows the analysis to be opened up in three important directions: first, it draws attention to the ways governing is assembled and mobilized; second, in dialogue with UPE it recognizes the centrality of material and semiotic circulations in understanding the governmental will to improve; and third, it asks us to pay more careful attention to understanding how and with what consequences new forms of climate subject are being created. In the remainder of this section we consider each of these avenues in turn.

A governmentality perspective emphasizes the ways in which the identification of issues which require some form of intervention do not neutrally appear, but rather are constituted through rationalities and techniques of governing such that imagining and defining the field of intervention is itself part of the process of governing. As Lockwood and Davidson (2010: 391) explain:

> a regime of practices … is constituted in response to the identification of an *object* that is problematized as needing governance. Objects such as 'nature' and 'natural resources' are identified as requiring intervention through processes that constitute the object itself – how it is understood, delineated and related to other objects …

How problems come to be understood as requiring some form of intervention is subject to broad rationalities, such as those associated with neoliberalism or social justice, as well as more specific discourses that are embodied in particular interventions or programmes (Foucault 2000d). Governmentality scholarship has tended to focus on how forms of neoliberal rationality have come to underpin the ways in which specific problems are conceived and addressed (Walters 2012: 71). However, this scholarship has often focused on neoliberalism as an overarching logic through which interventions are circumscribed, rather than being concerned with the specific means through which problematization takes place. Even in more nuanced accounts of the process and practice of problematization, the focus often remains on the discourses and rationalities at work, at the expense of the practices through which the will to improve is translated into specific forms of intervention (Death 2014 Li 2007b; Walters 2012). In order to move beyond this focus on the ways in which particular problems are framed and intervention sought, Li argues (2007b: 264) for an 'analytic of assemblage': a means through which to examine the mobilization of the heterogeneous elements in a regime of practices as it is 'assembled to address an "urgent need" and invested with strategic purpose'. As she stresses, particular programmes of intervention are 'not invented ab initio' but take place within an heterogeneous assemblage or *dispositif* such that they 'are pulled together from an existing repertoire, a matter of habit, accretion, and bricolage' (Li 2007a: 6). In her six-part framework, Li describes a set of practices that follow problematization: forging alignment, enrolling parties within the assemblage; rendering technical in which a field of intervention is delimited, a problem defined made visible and amenable to intervention; authorizing knowledge, where relevant bodies of expertise and evidence are identified and critiques contained; managing failures and contradictions, so that emerging tensions are regarded as solvable and compromises established; anti-politics, in which political questions are reframed as matters of technique and management; and reassembling, in which the assemblage is reworked as additional elements are added and existing elements reconfigured (Li 2007a: 7; Li 2007b: 265).

By demonstrating the intricate tactics and techniques through which the will to improve is assembled and mobilized, Li's approach highlights the importance of moving beyond an account focused only on the rationalities of government and attending instead to the ways in which power and politics is constituted and contested through the socio-materiality of the field of intervention itself. However, for the purposes of engaging with urban climate change experimentation we find this approach limited in two important ways. First, the focus on the technical (e.g. forms of calculation, expertise, management) and the mundane means through which this is achieved provides little scope for considering how interventions are animated by concerns with novelty and innovation or of the ways in which they build on compelling accounts through modalities of persuasion, seduction or inducement (Allen 2003). Second, the account offers little insight on how the will to improve is sustained over time. The rather linear narrative established by Li suggests that following contestation, governmental assemblages or regimes of practice are renewed and reconfigured. Our concern with the urban and with infrastructure systems therein leads us to ask how such projects endure, as well as to attend to the processes through which they break down and are reconfigured.

A second important avenue for analysis opened up by scholarship on governmentality lies in the centrality of circulations as both a component and outcome of government. Foucault (2009: 20) considered the field of intervention a milieu 'in which a series of uncertain elements unfold'. The milieu is 'what is needed to account for action at a distance of one body on another. It is therefore the medium of an action and the element in which it circulates' (Foucault 2009: 20–1). Regarded in this way, the regulation of the milieu as a political space 'involved not so much establishing limits and frontiers, or fixing locations, as, above all and essentially, making possible, guaranteeing, and ensuring *circulations*: the circulation of people, merchandise, and air etc.' (Foucault 2009: 29, emphasis added). In linking the improvement of the population to these forms of circulation, Foucault argues that 'the constitution of a knowledge (savoir) of government [or rationality] is absolutely inseparable from the constitution of a knowledge of all of the processes revolving around population in the wider sense of what we now call "the economy"' (Foucault 2009: 106). In short, government, as a form of power, is exercised in the maintenance of forms of circulation required for the operation of economies, be they those of the household, the flock or the state.

In recognizing the relationship between government, circulation, knowledge and economy, Foucault's work can be brought into dialogue with the field of UPE.

A diverse body of work, UPE scholars nevertheless share a common concern with the processes of socio-spatial transformation emerging from urbanization in the context of global capitalism and their social and environmental consequences (Heynen *et al.* 2006). UPE scholarship conceives of cities as 'dense networks of interwoven sociospatial processes that are simultaneously local and global, human and physical, cultural and organic' (Swyngedouw and Heynen 2003: 899), produced and reproduced through continual processes of circulation and transformation. Within UPE, circulation and transformation are regarded as *metabolic* processes, involving both human and non-human elements, and as the means through which capital accumulation takes place (Gandy 2005; Castán Broto *et al.* 2012). The apparent fixity and obduracy of the city is attained through the continual circulation of particular socio-environmental assemblages, and considerable work is required in order to ensure the continued dominance of particular urban forms. Within any particular locale, the history of socio-technical progress involves struggles for dominance and the reconfiguration of the technologies of power (Swyngedouw 1997). Engaging with social studies of science, and in particular actor network theory (ANT), UPE research conceives of the material and natural components of the urban as producing a cyborg urbanism (Gandy 2005). In doing so, UPE highlights the dynamic, unruly character of urban natures and materialities. Rivers, rats, invasive species, weather events, sewage, green spaces and so on profoundly shape both the physical and imagined spaces of the city (Desfor and Keil 2004; Kaika 2005).

Viewed in this manner, the forms of urban circulation which feature in Foucault's (2009: 18–20) development of the notions of security and governmentality are akin to the processes of urban metabolism that lie at the heart of the insights derived from UPE. Indeed, Foucault's 'example of the town' illustrates that governmentality was not only 'exercised over a whole population' (2009: 11) but also relies on 'a number of material givens' that work 'on site with the flows of water, islands, air and so forth' (2009: 19). As a milieu, the town:

> appears as a field of intervention in which, instead of affecting individuals as a set of legal subjects capable of voluntary actions ... one tries to affect, precisely, a population ... a multiplicity of individuals who are and fundamentally and essentially only exist biologically bound to the materiality within which they live.
>
> *(Foucault 2009: 21).*

The practice of governmentality is therefore 'a matter of maximizing the positive elements, for which one provides the best possible circulation, and of minimizing what is risky or inconvenient ... while knowing that they will never be completely suppressed' (Foucault 2009: 19). This in turn means that governmentality 'works on the future ... that is not exactly controllable, not precisely measured or measureable' (Foucault 2009: 20).

We suggest that critical insights for analysing experimentation as a form of urban climate governance can be garnered at the confluence of Foucault's work on government as a mode of power essentially concerned with circulation, his exploration of the means through which this takes place through the town, and contemporary UPE scholarship which has attended to the dynamics of such processes. We suggest that such a perspective entails attending to the ways in which experiments serve to reproduce or reconfigure different forms of urban circulation as critical in understanding their politics. Further, this approach reinforces the notion that the urban governance of climate change cannot be contained to specific sites and scales, but that it involves the constitution of particular fields of intervention through processes of circulation. As Swyngedouw and Heynen (2003: 912) argue, since urban dynamics are 'embedded within networked or territorial scalar configurations that extend from the local milieu to global relations', the focus of analysis needs to include 'the socioecological process through which particular social and environmental scales become constituted and subsequently reconstituted'. This in turn means that any governmental dispositif or regime of practices assembled in one urban site can take effect elsewhere, particularly considering the increasing demand for resources in urban areas and the structural inequality between the core and peripheries of urban regions (Castán Broto *et al.* 2012).

A third critical avenue for inquiry into the urban politics of climate governance that governmentality opens up concerns how and with what consequences new forms of climate subject are being created (Paterson and Stripple 2010). Subjects are seen not as passive agents, but as actively created as part of the constitution of the field of intervention and the will to improve. From a governmentality perspective, governing is achieved not through direct forms of order and control, but through *disposition*: 'arranging things so that people, following only their own self-interest, *will do as they ought*' (Li 2007a: 5). Subjects – that is, 'individuals and their institutions' – are considered to be 'both potentially governable through agency of their responses to direction, as well as being capable of thinking and acting in a manner contrary to that being sought by the governors' (Lockwood and Davidson 2010: 390). Rather than being subject to coercion or discipline, the practice of governmentality relies on the condition of freedom 'in which several kinds of conduct' are possible (Foucault 2000a: 342). While attention is frequently placed on the ways in which governmentality is enacted at a distance through techniques that aim to pass responsibilities to individuals and communities, Foucault's conception of subjectivity makes space for practices of resistance and subversion. Other traditions of critical social and political theory also serve to emphasize the means through which everyday practices can provide space for alternative dispositions and forms of conduct. In his critique of the 'Concept-city' (the city as produced by its rational organization), De Certeau (2002 [1984]) focuses on how everyday urban life challenges traditional conceptualizations of the city as a universal and anonymous subject. Instead, he contends, the urban is produced in city-practices whether it is walking, interpreting, inhabiting and exchanging the city:

> I would like to follow out a few of these multiform, resistant, tricky and stubborn procedures that elude discipline without being outside the field in

> which it is exercised, and which should lead us to a theory of everyday practices, of lived space, of the disquieting familiarity of the city.
>
> *(DeCerteau 2002 [1984]: 96)*

The city can thus only be governed by living it, because of the ways in which socio-technical systems are embodied and made real through daily practices and experiences. For example, the living experiences of Manhattan or Paris demonstrate that the city is not only a site of representation: it only becomes visible by walking through it (Latour and Hermant 1998; DeCerteau 2002 [1984]).

For all of its possibilities, scholars of governmentality have tended to use it to explain the discourses and rationalities that underpin particular policies or programmes. Adopting a programmer's view (Death 2014), governmentality perspectives have been found wanting in terms of their ability to engage precisely with that they claim to address – *how* governing takes place (Li 2007b). Thus, we adopt an alternative approach to understanding governmentality, one which recognizes that government is directed to a complex of 'humans and things', and where the practice of government involves not only the development of particular rationalities but, equally, the deployment of a suite of tactics and techniques. Our intention here is not to seek one true theoretical meaning in the extensive and diverse body of work on the notion of governmentality, but rather to engage in a critical encounter with these diverse strands of thought and analytical tools, in order to 'open up new ways of understanding social and political problems … and to revise and amend existing concepts' (Walters 2012: 5). Starting from such a position, we argue that attention must be paid to the practices of assembling programmes of intervention as a means through which to pursue the will to improve, to the 'hard work required to draw heterogeneous elements together, forge connections between them and sustain these connections in the face of tension … and [how they] might or might not be made to cohere' (Li 2007b: 264). Below, we develop such an approach further, tracing how governance extends from deliberate policy to alternative sites and spaces in the city, and arguing that what is required is attention to how forms of governmental intervention are *made*, *maintained*, and *lived*.

Experimentation and the art of government

We have argued above that the body of work surrounding the concept of governmentality enables us to move beyond existing conceptual approaches for understanding the governing of climate change and to engage in new critical encounters with how climate change has come to be governed in particular urban contexts. As Walters (2012: 20) suggests, Foucault's concern with the 'art' of government was first and foremost one of understanding 'how at different times, in specific places, and always in connection with particular political issues, certain experts, authorities, critics and dissidents have come to reflect on the problem of *how* to govern the state'. Examining the art of government enables analyses of how it is that climate change has come to be understood and responded to as a matter requiring government, and the ways in which experimentation has come to be deployed in this regard. Following the new avenues of enquiry discussed above, we find three particular apertures promising in relation to understanding how, why and with what consequences experimentation is becoming part of the art of urban climate change government. First, we suggest that understanding the emergence of experimentation requires engaging with the means through which particular fields of intervention come to be assembled, imbued with particular wills to improve and made practical through a variety of tactics and techniques so as to create particular forms of urban climate governmentality. Second, we suggest that attending to the ways in which the politics and practice of urban climate change governmentalities are bound up with the circulations of the economy as understood in its broadest terms (natural, physical, social) can enable a greater understanding of the nature of urban climate change experimentation. Third, we argue that examining the ways in which particular forms of subjectivity and conduct are bought into being, contested or resisted through forms of experimentation and in relation to particular urban climate change governmentalities is critical for understanding their politics and consequences.

From this basis, we propose a three-fold analytical framework which enables us to explore and explain the ways in which experimentation is becoming central to the art of urban climate change government. First, we argue that the *making* of urban climate change experiments is itself bound up in conceiving, framing and operationalizing the will to improve in particular urban milieu in relation to climate change. Second, we argue that experimentation involves a series of practices concerned with *maintaining* specific interventions, requiring: (1) the upkeep of their experimental qualities and the keeping at bay of unruly elements; and (2) a process of metabolic adjustment through which they are embedded in processes of circulation and urban reconfiguration. Third, we argue that experimentation involves techniques and tactics for ensuring experiments are *lived*. This involves engaging with the ways in which experiments intervene in the conduct of everyday life and involve the formation of climate subjectivities, as well as the ways in which they are resisted, circumvented and ignored.

Making experiments

How and why forms of experimentation come into being has been the key focus of the literature on the emergence of niches within socio-technical systems as the seedbeds for innovation and system change (see Chapter 1 and Chapter 12). Within this literature, niches are regarded as protected spaces, arenas within which technical and social innovations may be put to the test while remaining sheltered from wider economic and political forces. While often regarded in highly technical terms, some authors working in this domain have pointed to the important social processes involved in fostering niches and experiments, including the development of specific ideas, the alignment of multiple actors, and the bringing together of different social and material constituents to form an experimental space (Smith *et al.* 2010). The key aspect of this literature is the recognition of situated interventions as embodying the potential for wider structural change. However, as with the broader field of transition studies within which they are located, these analyses lack an explicit conceptualization of power (Avelino and Rotmans 2011). The interactions between growing niches and dominant regimes points towards a struggle of power, but how this is played out in practice is rarely analysed. Avelino and Rotmans (2011) have developed a framework based upon Foucault's theorization of power that relates struggles over meaning and relations to structural change. In contrast, the concept of experiment we deploy here directs attention away from the relationship between niche and regime towards the on-going translations and assemblages of multiple material and non-material elements through which politics are exercised and governmental projects conducted (Bulkeley and Castán Broto 2013). By specifically focusing on the practices, techniques and arts of governmentality it is possible to examine the ways in which experiments are deliberately built upon both tactics and strategies that belong simultaneously to niches and regimes (cf. Lovell *et al.* 2009).

Rather than starting from the perspective of niches within wider socio-technical regions, approaching experimentation through the analytics of assemblage provides an alternative starting point from which to consider how and why experimentation takes place. If, as argued in the previous section, government is the directed pursuit of a will to improve, then it follows that it is of critical importance to understand how fields of intervention come to be constituted through a range of tactics, including problematization, rendering technical, assembling and authorizing knowledge, representing the field of intervention in particular ways, and closing down political dissent (Lockwood and Davidson 2010: 391; Li 2007b). As Li argues:

> to render a set of processes technical and improvable an arena of intervention must be bounded, mapped, characterized, and documented; the relevant forces and relations must be identified; and a narrative must be devised connecting the proposed intervention to the problem it will solve.
>
> *(Li 2007a: 126)*

These processes are necessarily intimately connected, since the delimitation of one depends on the understanding of the other. Problems do not come fully formed, but are shaped by the solutions at hand, and solutions confer authority and agency on those who would intervene in relation to any particular problem (Li 2007; Hajer 1995; Owens and Cowell 2002). This connection between the rationalities at work in the process of problematization and the techniques of the practices of rendering technical is critical to the way in which the focus of governmental will comes to be determined.

Such insights suggest that central to the making of experiments are the ways in which climate change comes to be understood as an urban issue through a process of problematization. To govern climate change in the city is to address questions of risk and security in environmental, economic and political terms (Hodson and Marvin 2009; While *et al.* 2010). At the same time, work on urban responses to climate change is replete with references to opportunities and the potential to create alternative futures. Experimentation may provide a means through which such apparent disjunctures can be resolved. Such problematizations give rise to and are further configured through practices of assembling a programme of improvement through which they can be addressed. This occurs through what Li (2007b: 265) terms 'forging alignment', bringing into productive relation the different parties and, we would add, the material elements that constitute the field of intervention. This resonates with the insistence of scholars within science and technology studies that the ways in which constituent elements of such assemblages are bought into alignment serves to sustain particular orders or regimes (Blok 2012). Forging alignments requires tactics and techniques of rendering technical, of creating the 'limits and particular characteristics' of the field to be governed through multiple techniques of calculation, visibility, generating information and developing practical interventions (Li 2007a: 7). In the case of experimentation, however, we suggest that such tactics are particularly significant in creating *spaces of exception* as the delimited arenas for intervention. At the same time, making such interventions practical and workable is critical, in order to translate them from mere aspirations into doable projects – techniques of calculation, visibility, evidencing and audit are particularly useful in this regard. Thus the making of experiments relies both on opportunities to ensure that experimentation is visible and the ability of intervening agencies to enrol actors, materials and resources through the production of suitable narratives that legitimate the experiment as a form of intervention.

In addition to working at the cusp of what is innovative and that which can be counted upon to deliver, we suggest that experimentation, as an art of government, also operates by simultaneously establishing and challenging the boundaries of the field of intervention. Making urban climate change experiments is dependent on rendering climate change technical – known, calculated and represented – within a confined urban locale. However, rather than regarding government as a process through which the boundaries of intervention are settled, we suggest that experimentation is precisely about boundary testing – of feeling where and how the field of intervention for addressing climate change might be circumscribed, and of exploring whether such interventions do indeed lead to improvement for the multiple 'specific finalities' (Foucault 2009: 99) that they seek to address. Rather than working with a degree of assurance that 'calculated interventions will produce beneficial results' (Li 2007b: 270), we suggest that the making of experiments is undertaken as a means of seeking to test and experience what it means to respond to climate change, and of establishing what improvement entails. In doing so, experiments do not only test the boundaries of what it means to address climate change the city, but also probe the nature of the city itself and its spatial, material and social limits. Thus, our analysis of experimentation in the making draws attention to the ways in which the art of government is concerned with the unfolding in time and space of uncertain elements, in which the work involved in addressing climate change is contingent and bounded to a particular urban locale, yet open to multiple possibilities.

Furthermore, we find that a further set of practices is essential to making experiments and mobilizing the will to improve, in part due to the provisional nature of experimentation as an art of government: there must be some means through which such interventions are made compelling. The novelty and risk implied by experimentation on the one hand, and its use as a means of testing out different alternatives on the other, requires that those who would govern are able to persuade, cajole and induce others to participate in that which is uncertain and where the rewards are unclear. Rather than rendering interventions technical as such, this suggests that there is considerable emotional and affective work to be done in the making of experimentation, and points to a different set of techniques, including publicizing, networking and championing, which are also important in this form of government. The *making* of climate change experiments, in our view, involves at least four distinct but interdependent components: problematization, alignment, rendering technical and rendering compelling.

Maintaining experiments

Moving beyond an understanding of how experiments come to be assembled and pursued requires, we suggest, a greater engagement with the ways in which they unfold within the urban milieu (Foucault 2009: 20). Alongside an understanding of the practices of assemblage and enclosure involved in the making of particular fields of intervention, there is also a need to attend to the ways in which experiments are engaged in the workings of different forms of circulation and the extent that they involve 'organizing, or anyway allowing the development of ever-wider circuits' (Foucault 2009: 45). These practices, we argue, can be regarded as forming the work of *maintaining* experiments. Drawing specifically on insights from urban political ecology, we argue that two distinct but related forms of practice are central to the maintenance of experiments: *upkeep* and *metabolic adjustment* (Castán Broto and Bulkeley 2013).

In relation to the first set of practices, maintenance entails processes of daily upkeep, the crucial repair work through which the city is sustained (Graham and Thrift 2007). Through these sets of processes, the material and social fabric is maintained and improved and the appearance of particular facets of the city made to endure. As Graham and Thrift (2007) point out, upkeep involves mundane practices (washing, scrubbing, painting, removing waste, freeing drains), as well as more strategic interventions (the closure of highways for road repairs, investment in new plant required to run existing energy networks). Upkeep both sustains particular infrastructures in the face of continual processes of decay and enables them to keep up appearances. In the case of experimentation, upkeep is required both in order to maintain the physical and technical features of innovation in working order by embedding interventions in particular contexts and ensuring their smooth operation, but also to ensure that the notion of the intervention as experimental is upheld, made visible and circulated. Furthermore, such practices serve not only to sustain the experiment in its original form, but also provide 'a vital source of variation, improvisation and innovation' (Graham and Thrift 2007: 6).

In contrast to upkeep, metabolic adjustment relates to the ways in which the insertion of experiments in particular urban milieu constitute attempts to reconfigure existing forms of circulation. Dynamic circulation makes possible the 'solidification' of existing economic and political structures (Harvey 1996). The circulation of materials, capital and organizing principles serves to convey the permanence and obduracy in the city (Star 1999; Hommels 2005). At the same time, the dynamic and emergent nature of such processes means that they cannot easily be domesticated to serve instrumental processes: urban materialities are characterized by being unruly (Kaika 2005). Maintenance, from this perspective, is the set of practices that enables adaptation to the dynamics of urban circulation, 'a broadly structural process of achieving or containing particular forms of metabolic circulation' (Castán Broto and Bulkeley 2013: 1936). Furthermore, 'orchestrated by forms of connection, disconnection, and fragmentation … maintenance becomes critical to the reproduction of the material/infrastructural in everyday life, and to its continuing structuring effects in the city' (Castán Broto and Bulkeley 2013: 1936). Yet, the practice of metabolic adjustment may also be directed to new ends. Policy mobilities, technological developments or the development of new cultural sensibilities, for example, may lead to deliberate attempts to disrupt existing urban infrastructure lattices as a means to adapt to new urban contexts (Chapter 1). This process also requires experiments to facilitate alternative forms of metabolic adjustment. In the case of climate change, the emergence of new carbon calculus or low carbon urbanism (While *et al.* 2010) and new demands from urban publics may foster the need for new forms of experimentation which enable the readjustment of urban dynamics.

Situating the interventions through which governmental programmes are realized within the wider context of Foucault's theorization of government as the means through which improvement was sought through the operation of 'economy', we argue that the study of the practice of government needs to attend to the ways in which these forms of circulation are maintained. The concept of maintenance provides a useful aperture through which to view these processes, enabling us to attend to both the structural, purposive ways in which experiments intervene to reconfigure network circulation through metabolic adjustment, and to the micro-practices through which the 'upkeep' of experiments is achieved. We argue that:

> upkeep and metabolic adjustment processes have a dual function in the sense that they are simultaneously directed towards ensuring the permanence of the experiment within the given infrastructure context, and asserting its potential for provoking change … [so that] maintenance is required both to sustain the novelty of experiments and to normalize them within particular urban contexts.
>
> *(Castán Broto and Bulkeley 2013: 1938)*

Such practices work both to authorize knowledge, demonstrating how experimental interventions have succeeded, and as a means through which the techniques, calculations and representations involved in making experiments come to have a tangible presence which is, in turn, capable of being extended and mobilized beyond the specific site of intervention. Practices of maintenance are also a means through which the excesses of innovation, the unruly metabolisms of cities, and the failures and contradictions inherent in seeking to accomplish the will to improve (Li 2007b) are managed. Innovations break down, efficiencies in energy use fail to materialize, and techniques of various kinds come unstuck. Practices of maintenance enable the continual reworking of what it is that the experiment is concerned with, and provide both for the striking of new alignments and compromises, and the admission of failure. By attending to the ways in which such practices are materially undertaken, and their insertion within wider networks and processes of urban metabolism, this analytical device allows us to extend and deepen our understanding of the practices through which governing is conducted.

Living experiments

A third set of insights that we draw upon in seeking to understand the work of experimentation are those that examine the ways in which government as the conduct of conduct rests upon establishing particular dispositions and subjectivities. In this sense, experimentation, to take effect, has to be *lived* – taken up in the day-to-day practices of the individuals and institutions that are subject to forms of governmental intervention. Subjectivities are formed through the direction of particular rationalities towards specific ends (Lockwood and Davidson 2010: 390). To this end, Foucault emphasizes the importance of the ways in which conduct comes to be arranged (*disposer*) for 'it is not a matter of imposing a law on men, but of the disposition of things, that is to say, of employing *tactics* rather than laws' (Foucault 2009: 99, emphasis in original). Analyses of neoliberal governmentalities have emphasized the ways in which such rationalities are predicated on governing through the self or governing by community (Paterson and Stripple 2010). However, 'the dominant tendency to focus on governmental rationalities' within this approach 'contributes to a view of power as top down, totalizing and omnipresent', and often neglecting the specific techniques through which such subjectivities are imagined, directed, resisted and contested in the messy actualities of governmental practice (McKee 2011: 3). Experiments operate in uncertain terrains. Totalizing rationalities and alternative subjectivities are under negotiation. Both making and maintaining experiments requires a degree of flexibility and risk taking at every step. Assemblages are often improvised, built upon existing local resources and social networks. Uncanny natures are kept at bay through endless repetition of upkeep routines. The continual adjustment of the experiment as it is maintained also requires that subjectivities are kept open, and conduct improvised, suggesting in turn that these are far from fully formed visions or a dominating force.

In addition to the specific forms of conduct and subjectivity that experimental interventions seek to foster, we suggest that experimentation also involves the creation of affective links across the multiple elements that are assembled in the experiment, creating a form of 'experimental subject'. Those subjects which are both party to its governance and who are governed through the experiment develop their own commitment towards it as it is realized, creating new forms of interdependency and a degree of shared senses of purpose and investment. This necessarily requires a degree of mutual commitment if experimental rationalities are to be incorporated into the affective aspects of creating new subjects. Contestation and resistance to these commitments – and their subsequent negotiation – is thus necessary for experiments to become lived. Thus, the tactics and techniques of living experiments involves the means through which the right disposition of things for particular forms of conduct is achieved, together with the ways in which this is contested in often mundane and incremental ways.

In this way, experimentation as lived practice, conduct and subjectivity also involves encountering the limits of governmentality. On the one hand, forms of subjectivity and conduct central to the working of experimentation may be contested or simply neglected. On the other, the interdependences created through belonging to the experiment necessarily entail compromise, which can both confine what it is that the experiment can achieve and lead to significant tensions between the parties involved. In each case, the techniques and practices of living experiments work across the boundaries between the exceptional and the ordinary, the technical and the political, requiring constant work to both police the boundary and to hold dissent at bay. As a result, by focusing on everyday practice within urban climate change experiments we are able to draw attention to the contingent nature of socio-technical intervention. Experiments, as sites of intervention and contestation, are also a means through which the technical and political domains of government are blurred and bought into question. In her analysis of the practice of government, Li argues that rendering technical simultaneously renders non-political, such that questions of, in this case, the causes and consequences of both climate change and its response are side-lined in favour of some form of resolution. There is, however, no guarantee that this boundary can be maintained, for 'if appropriate subjects could not be identified, or their activities did not conform to requirements, there would be further compromises, instabilities and reversals' (Li 2007a: 199).

Conclusion: climate change experimentation and the politics of the city

In this chapter, we have sought to develop a new conceptual perspective through which to analyse the ways in which experimentation has come to form part of the means through which the governing of climate change is taking place in different urban milieu. Rather than regarding such interventions as purely technical endeavours or isolated examples of policy best practice, we have sought to develop a perspective that can engage critically with why experimentation has come to be part of the art of governing climate change and how this is being undertaken in different contexts. In so doing, we have drawn on the broad corpus of work that has been established around Foucault's concept of governmentality. Rather than offering an established theory or coherent perspective, governmentality has been likened to a toolbox, a 'cluster of concepts that can be used to enhance the thinkability and criticizeability of past and present forms of governance' (Walters 2012: 2) and which is most productively used not in wholesale application, but rather in the form of a 'critical encounter' (Walters 2012: 4). In this spirit, we have sought to draw on insights from those who have developed an account of the governmentality as a set of processes, tactics and techniques, assembled and contested in the practice of politics and everyday life, and engage these with the work from UPE and science and technology studies which has sought to understand the political economies and ecologies of cities and infrastructure systems. The result is not intended to be a comprehensive theory through which to comprehend the totality of urban responses to climate change, but rather a vantage point from which to examine their myriad manifestations.

From this starting point, we argue that experimentation is a central means through which the 'art of urban climate change government' is being conducted, and a crucial means to develop tactics and strategies through which climate change is being problematized in diverse urban settings, ranging from the broad discourses that inform policy to the situated practices of everyday life. Experimentation is an ongoing process in which successive stages of rendering the climate change problem technical, situating it within a locale, enrolling agents and discourses, and engaging in contestation may happen simultaneously. It is a collective effort to test the limits of governing climate change – of its potential, and its finalities.

We suggest that experimentation can be conceived productively as constituted through three distinct processes: making, maintaining and living (see Table 2.1). These processes highlight the fact that experiments are ambiguous, both challenging incumbent socio-technical assemblages and contributing to their reproduction. The making of experiments is the process in which socio-technical assemblages are constructed in response to specific problematizations of climate change in urban contexts, through novel alignments which constitute new fields of intervention. In this, the process of experiment-making resonates with the literature on policy mobilities, as it is directed towards the creation of sites of intervention, and with that of socio-technical regimes that regard experiments as part of the development

TABLE 2.1 Key processes of the Making-Maintaining-Living conceptual framework

Making	*Maintaining*	*Living*
• Problematisation • Alignment • Rendering technical • Rendering compelling	• Upkeep • Metabolic adjustment	• Subjectification • Contestation • Resistance

of specific niches (Chapter 1). Moreover, given the conceptualization of climate change as a global problem with local consequences, making experiments is an attempt to negotiate the perceived scale divide between the unfolding of the problem and the possibilities of intervention. In doing so, the emergence of experimentation can in part be explained by the global context of authority fragmentation, the growing policy vacuum surrounding global responses to climate change and the proliferation of *ad hoc* action following the discontent with global politics (Hoffmann 2011). In dialogue with this perspective, our analysis argues that experimentation is emerging as a particular form of urban politics in the context of low carbon urbanism which forms the milieu (Bulkeley *et al.* 2010; Hodson and Marvin 2010; While *et al.* 2010). We suggest that processes of rendering climate change technical and compelling will be critical to the ways in which the making of climate change experimentation takes place.

The second strand of our analytical approach concerns the process of maintaining experiments. Maintaining explicitly relates our analysis of experimentation to the notion of material and semiotic circulations which are central in the production of the urban. This concern is of critical importance to the work on UPE and also features in the development of Foucault's work on the exercise of power as government. In the context of urban circulations, we suggest that experiments at once seek integration within existing urban circulations while at the same time challenging the socio natural assemblages within which they emerge. In this sense, maintaining works to both ensure the normality of the experiment and to sustain its exceptional nature (Chapter 11). Maintaining is also a crucial process for understanding the impacts and implications of experiments, and how they travel (Chapter 12). At issue is not the scaling up (or down) of experiments, but rather the means through which elements from particular forms of experiment come to be integrated and flow through urban circulations on the one hand, and how these different circulations are reconfigured (or not) because of the presence of experimentation with its potential for replication.

Finally, we identify living as a process that relates experiments to the production and contestation of new subjectivities. Experiments come to be lived through the creation and contestation of subjectivities (in place and elsewhere), and by continually exceeding the boundaries placed around them. They raise questions of politics, not politics writ large, but rather politics writ small, emerging within an urban context through the capillary-like distribution of power that everyday interactions and circulations provide. In doing so, our analysis of experiments moves away from other Foucauldian debates on the anti-politics of intervention and the constitution of political decisions because these are expert-based (Ferguson 1990). It also raises questions about the assertion that climate change can be understood as a post political issue, governed through global or local consensus (Swyngedouw 2010). Instead, it suggests that experiments are intrinsically political. Experiments may appear as strategic attempts to reconfigure power in an uncertain context by dominant socio-economic and political actors. Yet these strategic attempts are led by various actors and the results are both unknown and incalculable. As Walters

(2012: 81) suggests 'the lines, contexts and energies of politicization can seldom be easily predicted in advance', such that there is no simple equation to be reached between the exercise of governmental power and practices of de-politicization. Indeed, as we explore throughout the cases in this book (and particularly in Chapter 11), at the heart of climate change experiments are fundamental questions of social and environmental justice, expressed in terms of rights, responsibilities and the recognition of inequalities, which emerge in multiple and often unpredictable ways to shape the urban politics of climate change.

References

Allen, J. (2003) *Lost Geographies of Power*. Oxford: Blackwell.

Avelino, F. and Rotmans, J. (2011) 'A dynamic conceptualization of power for sustainability research'. *Journal of Cleaner Production*, 19(8): 796–804.

Betsill, M. M. and Bulkeley, H. (2006) 'Cities and the multilevel governance of global climate change'. *Global Governance*, 12(2): 141–59.

Biermann, F. and Pattberg, P. (2008) 'Global environmental governance: taking stock, moving forward'. *Annual Review of Environment and Resources*, 33(1): 277–94.

Blok, A. (2012) 'Greening cosmopolitan urbanism? On the transnational mobility of low-carbon formats in Northern European and East Asian cities'. *Environment and Planning A*, 44(10): 2327–43.

Bulkeley, H. and Newell, P. (2010) *Governing Climate Change*. London: Routledge.

Bulkeley, H. and Castán Broto, V. (2013) 'Government by experiment? Global cities and the governing of climate change'. *Transactions of the Institute of British Geographers*, 38: 361–75.

Bulkeley, H., Castán Broto, V., Hodson, M. and Marvin, S. (eds.) (2010) *Cities and Low Carbon Transitions*. Abingdon: Routledge.

Burch, S. (2010) 'Transforming barriers into enablers of action on climate change: insights from three municipal case studies in British Columbia, Canada'. *Global Environmental Change*, 20(2): 287–97.

Castán Broto, V. and Bulkeley, H. (2013) 'Maintaining climate change experiments: urban political ecology and the everyday reconfiguration of urban infrastructure'. *International Journal of Urban and Regional Research*, 37(6): 1934–48.

Castán Broto, V., Allen, A., & Rapoport, E. (2012) 'Interdisciplinary perspectives on urban metabolism'. *Journal of Industrial Ecology*, 16(6), 851–61.

Chu, S. Y. and Schroeder, H. (2010) 'Private governance of climate change in Hong Kong: an analysis of drivers and barriers to corporate action'. *Asian Studies Review*, 34(3): 287–308.

Coafee, J. and Healey, P. (2003) 'My voice: my place: tracking transformations in urban governance'. *Urban Studies*, 40(10): 1979–99.

Dean, M. (2007) *Governing Societies: political perspectives on domestic and international rule*. Maidenhead: Open University Press.

Death, C. (2014) 'The limits of climate governmentality'. In Stripple, J. and Bulkeley, H. (eds) *Governing the Climate: new approaches to rationality, power and politics*. New York: Cambridge University Press, pp. 77–91.

De Certeau, M. (2002 [1984]) *The Practice of Everyday Life*. Trans. Steven Rendall. Berkeley: University of California Press.

Desfor, G. and Keil, R. (2004) *Nature and the City: making environmental policy in Toronto and Los Angeles*. Tucson, AZ: University of Arizona Press.

Dingwerth, K. and Pattberg, P. (2006) 'Global governance as a perspective on world politics'. *Global Governance: A Review of Multilateralism and International Organizations*, 12(2): 185–203.

Ferguson, J. (1990) *The Anti-Politics Machine: 'development', depoliticisation, and bureaucratic power in Lesotho*. Cambridge: Cambridge University Press.

Foucault, M. (2000a) 'The Subject and Power'. In Faubion, J. (ed.) *Power: essential works of Foucault 1954–1984*, Volume 3. New York, NY: New Press, pp. 326–48.

——(2000b) 'The Political Technology of Individuals'. In Faubion, J. (ed.) *Power: essential works of Foucault 1954–1984*, Volume 3. New York, NY: New Press, pp. 298–325.

——(2000c) 'Questions of Method'. In Faubion, J. (ed.) *Power: essential works of Foucault 1954–1984*, Volume 3. New York, NY: New Press, pp. 223–38.

——(2000d) '"Omnes et Singulatim": toward a critique of political reason'. In Faubion, J. (ed.) *Power: essential works of Foucault 1954–1984*, Volume 3. New York, NY: New Press, pp. 298–325.

——(2009) *Security, Territory, Population: lectures at the College de France, 1977–78*. Ed. M. Senellart, trans. G. Burchell. Basingstoke: Palgrave MacMillan.

Gandy, M. (2005) 'Cyborg urbanization: complexity and monstrosity in the contemporary city'. *International Journal of Urban and Regional Research*, 29(91): 26–49.

Graham, S. and Thrift, N. (2007) 'Out of order: understanding repair and maintenance'. *Theory, Culture and Society*, 24: 1–25.

Hajer, M. A. (1995) *The Politics of Environmental Discourse: ecological modernization and the policy process*. Oxford: Clarendon Press.

Harvey, D. (1996) *Justice, Nature and the Geography of Difference*. New York: John Wiley.

Heynen, N., Kaika, M. and Swyngedouw, E. (2006) 'Urban Political Ecology: politicizing the production of urban natures'. In Heynen, N., Kaika, M. and Swyngedouw, E. (eds) *In the Nature of Cities: urban political ecology and the politics of urban metabolism*. London: Routledge, London, pp. 1–20.

Hodson, M. and Marvin, S. (2009) '"Urban ecological security": a new urban paradigm?'. *International Journal of Urban and Regional Research*, 33(1): 193–215.

——(2010) 'Can cities shape socio-technical transitions and how would we know if they were?'. *Research Policy*, 39(4): 477–85.

Hoffman, M. (2011) *Climate Governance at the Crossroads: experimenting with a global response after Kyoto*. Oxford: Oxford University Press.

Hommels, A. (2005) 'Studying obduracy in the city: toward a productive fusion between technology studies and urban studies'. *Science, Technology and Human Values*, 30(3): 323–51.

Jagers, S.C. and Stripple, J. (2003) 'Climate governance beyond the state'. *Global Governance*, 9(3): 385–400.

Kaika, M. (2005) *City of Flows*. London: Routledge.

Latour, B. and Hermant, E. (1998) *Paris Ville Invisible*. Paris: Institut Synthelabo pour le progres de la connaissance.

Li, T. M. (2007a) *The Will to Improve: governmentality, development, and the practice of politics*. Durham, NC: Duke University Press.

——(2007b) 'Practices of assemblage and community forest management'. *Economy and Society*, 36(2): 263–93.

Lockwood, M. and Davidson, J. (2010) 'Environmental governance and the hybrid regime of Australian natural resource management'. *Geoforum*, 41(3): 388–98.

Lövbrand, E. and Stripple, J. (2012) 'Disrupting the public–private distinction: excavating the government of carbon markets post-Copenhagen'. *Environment and Planning C: Government and Policy*, 30(4): 658–74.

Lovell, H., Bulkeley, H. and Owens, S. (2009) 'Converging agendas? Energy and climate change policies in the UK'. *Environment and Planning C: Government and Policy*, 27(1): 90–109.
McKee, K. (2011) 'Sceptical, disorderly and paradoxical subjects: problematizing the "Will to Empower" in social housing governance'. *Housing, Theory and Society*, 28(1): 1–18.
Murdoch, J. (2000) 'Space against time: competing rationalities in planning for housing'. *Transactions of the Institute of British Geographers*, 25(4): 503–19.
Oels, A. (2005) 'Rendering climate change governable: from biopower to advanced liberal government?' *Journal of Environmental Policy and Planning*, 7(3): 185–207.
Okereke, C., Bulkeley, H. and Schroeder, H. (2009) 'Conceptualizing climate governance beyond the international regime'. *Global Environmental Politics*, 9(1): 58–78.
Owens, S. E. and Cowell, R. (2002) *Land and Limits: interpreting sustainability in the planning process*. London: Routledge.
Paterson, M. and Stripple, J. (2010) 'My Space: governing individuals' carbon emissions'. *Environment and Planning D: Society and Space*, 28(2): 341–62.
Risse, T. (2004) 'Global governance and communicative action'. *Government and Opposition*, 39(2): 288–313.
Schreurs, M. A. (2010) 'Multi–level governance and global climate change in East Asia'. *Asian Economic Policy Review*, 5(1): 88–105.
Sending, O. J. and Neumann, I. B. (2006) 'Governance to governmentality: analyzing NGOs, states, and power'. *International Studies Quarterly*, 50(3): 651–772.
Smith, A., Voß, J-P. and Grin, J. (2010) 'Innovation studies and sustainability transitions: the allure of the multi-level perspective and its challenges'. *Research Policy*, 39(4): 435–48.
Star, S. L. (1999) 'The ethnography of infrastructure'. *American Behavioural Scientist*, 43(3): 377–91.
Stripple, J. and Bulkeley, H. (2014) *Governing the Climate: new approaches to rationality, power and politics*. New York: Cambridge University Press.
Swyngedouw, E. (1997) 'Neither global nor local: "glocalization" and the politics of scale'. In Cox, K. (ed.) *Spaces of Globalization*. New York: Guilford Press, pp. 137–66.
——(2004) 'Globalisation or "glocalisation"? Networks, territories and rescaling'. *Cambridge International Review of International Affairs*, 17(1): 25–48.
——(2010) 'Apocalypse forever? Post-political populism and the spectre of climate change'. *Theory Culture & Society*, 27(2–3): 213–32.
Swyngedouw, E. and Heynen, N. C. (2003) 'Urban political ecology, justice and the politics of scale'. *Antipode*, 35(5): 898–918.
Walters, W. (2012) *Governmentality: critical encounters*. New York: Routledge.
While, A., Jonas, A. E. G. and Gibbs, D. (2010) 'From sustainable development to carbon control: eco-state restructuring and the politics of urban and regional development'. *Transactions of the Institute of British Geographers*, 35(1): 76–93.

PART II

3

TOWARDS ZERO CARBON HOUSING IN BANGALORE, INDIA

Introduction

Urbanization in India is not only a spatial phenomenon marked by the rapid growth of cities but also a wider cultural shift towards new ways of living. As urbanization continues apace, problems of infrastructure supply and energy and water shortage have become acute, yet at the same time the continuous increase in wealth in the country, albeit unevenly distributed, has led to noticeable changes in the cities of India. Reflecting these trends, urban planning models look towards the West, especially the USA, for points of reference, and returning highly-qualified émigrés settle in the Indian cities for lifestyles that resemble their experience of western lifestyles (e.g. Nair 2006). Bangalore, the capital of the state of Karnataka and the fifth largest city in India, epitomizes these urban dynamics and the challenges they pose. In 2006 the estimated population was 5,686 million, but at a growth rate of 2.79 per cent a year, Bangalore is expected to reach 10 million people by 2020. The city's highly ethnically diverse population is demonstrated by the everyday use of different languages, such as Kannada, Urdu, English, Hindi and Telugu. Bangalore's GDP has been estimated at US$69 billion and growing at a rate of 6.5 per cent per year. It is estimated to grow to US$203 billion by 2020, thus making Bangalore one of the richest cities in the world (PricewaterhouseCoopers 2009). This huge increase in both the population and wealth of the city has been driven by the growth of the IT and offshoring sectors since the 1990s.

Economic and population growth has placed unprecedented demands on Bangalore's infrastructure and generated growing urban inequalities. After decades of underinvestment, Bangalore's infrastructure is regarded as insufficient for the demands of the new economic players and for providing basic services to the city's residents. Part of the problem is that the rapid urbanization associated with economic growth in the city has disrupted the network of water tanks established during colonial times to guarantee the water supply to the city. The number of lakes has been reduced from 400 to 64 (Government of Karnataka 2007). Furthermore, the removal of green cover and well extraction threatens the capacity of the city's groundwater reservoirs to replenish themselves. Water scarcity means that water needs to be transported over long distances, increasing the energy consumption of the city (Mukhopadhyay and Revi 2009) and adding to the inequality in water provision across the city. The Bangalore Water and Sewerage Board (BWSSB) maintains a piped network that currently supplies 35 per cent of households (alternative estimations up to 60 per cent can be found). A proportion of these connections are supplied from the Cauvery River, but there is an overall reliance on groundwater resources (see Table 3.1). The BWSSB also maintains 5,850 bore wells and 15,180 public taps, which together supply a further 30 per cent of the city's needs (Government of Karnataka 2007). This does not include the inestimable number of private bore wells which are particularly associated with new developments. Climate change has become an important discourse linking the challenges of everyday life with the city's global aspirations, at the same time as small- and medium-sized enterprises have been threatened by the globalization of the city's economy, which has already led to increasing rates of poverty and inequality.

This chapter explores an experiment that has emerged as a response to both climate change and the urbanization of Bangalore: Towards Zero Carbon Development

TABLE 3.1 Modes of access to water provision by different groups of population in Bangalore

Mode of access	*Who accesses*	*Provider*
Handpumps (very few functioning)	Poor groups in urban villages and revenue layouts	Public, (Urban Local Bodies, ULB)
Borewell water (500–800 ft depth) stored in mini water tanks with attached public taps	Poor groups living in urban villages, slums and revenue layouts	Public (ULB)
Piped borewell water	Lower-middle income groups in revenue layouts	Public (ULB)
Tankers (sourced from private borewells belonging to large landowners)	Middle class households	Private
Individual borewells (800–1,200 ft in depth)	Wealthier middle class households. Sufficient land and documents are needed to get a power connection to pump the water up	Private
Bottled drinking water	Wealthier middle class households	Private
Piped water from the Cauvery river (2–4 hours per day, every other day)	10% of periphery in BDA-approved areas only (technology parks and few large apartment complexes etc.)	Public (Bangalore Water and Sewerage Board)

Source: Adapted from Ranganathan *et al.* (2009).

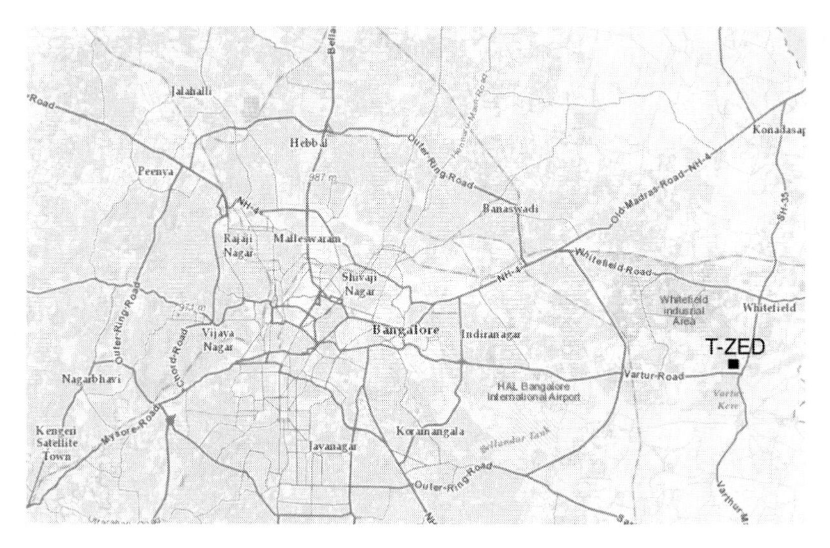

FIGURE 3.1 Map of Bangalore showing the location of T-Zed. Drawn by V. Castán Broto on ArcGIS Basemap.

(T-Zed). A gated community, T-Zed was intended to demonstrate the potential for environmentally sensitive living that did not compromise the new middle class' consumer expectations for luxury housing. T-Zed provides a means to render technical a range of housing technologies for low carbon development. Yet T-Zed it sits uncomfortably in the overall urban landscape of Bangalore (Figure 3.1). Maintaining and living T-Zed, beyond its initial assemblage, required the painstaking work of maintenance and repair which would have been impossible without the subjectification of individuals within Bangalore's middle classes who assumed responsibility for responding to climate change grounded in opportunities to buy it in the design of their houses and the control of certain aspects of their ways of living, without questioning broader inequalities across the city. The project's greatest success has been creating a model of the green consumer in Bangalore and establishing new justifications for existing models of development. Overall, we find that this form of experimentation succeeded in terms of delivering housing innovation for climate change, but failed to engage with the issues of social justice emerging at the boundaries of urban development in the city.

Towards zero carbon development: housing the emerging middle class in India

When T-Zed was completed in 2007, it was recognized by several developers and architects as a landmark project that had demonstrated what green development could bring to Indian cities. The project was developed by the private company Biodiversity Conservation India Limited (BCIL) as a model that could compete with regular urban homes. The central idea was to build a near-zero carbon development without compromising the habits and lifestyles of their target market, who were considered high-energy consumers. The two key narratives orchestrating T-Zed are that of carbon reduction and securitization of the city's resources. Whereas carbon reduction is the selling point of the development, intended to capture the imagination of future residents, its features are articulated in terms of the securitization of resources to sustain particular forms of individual lifestyle. BCIL estimated that, in comparison with a standard development supplied by piped water, T-Zed could save over 2 million rupees in ten years on water supply and four times that amount on energy supply by being self-sufficient, while also providing other significant benefits, such as improved health, fresh environments, clean air and enjoyment of surroundings.

However, while BCIL understandably emphasized the self-sufficiency of gated developments like T-Zed, such forms of resource securitization are also fundamental to facilitating the levels of consumption of energy and water expected by the residents of such developments. In the context of a lack of basic service provision across the city, gated compounds create access to lifestyles unthinkable in other parts of the city. Self-sufficiency, pursued here under the auspices of demonstrating the potential for zero carbon development, also provides a means through which to secure such forms of urban development. As intermediaries between customers confused over what sustainability might entail and limited in terms of their choices in the market, in their role as a building developer constructing future urban landscapes, BCIL has emphasized its role in innovation. T-Zed was regarded as the original field laboratory where BCIL ideas were first taken to a commercial scale, experimenting not only with concepts of building but also with new relationships between developers and customers. In this manner, innovation is perceived not as an isolated feature within the development, but as a disposition which pervades the development in both social and technical forms.

Technical innovation

In an attempt to govern climate change through building design, the group of idealist engineers and architects who were first drawn toward the T-Zed experiment worked towards finding ways to translate their understanding of climate change into possibilities for building design. In doing so, they drew both on their international experience and their own involvement in other pilot projects in Bangalore. The carrying capacity of the land was the central concept that shaped T-Zed's design. Through this notion, BCIL sought to intervene in the local debate concerning the relative merits of densification and urban sprawl by designing a development that was sufficiently compact to realize its environmental ambitions but which had enough land to secure its resource needs in terms of energy and water. In the context of Bangalore, the availability of both groundwater and rainfall were seen as the main limiting factors for the size of the development. On this basis, BCIL technicians established minimum requirements of water per family and the land required to ensure capture that volume of water. They calculated that their plot of 5.23 acres (21,160 m^2) could host a maximum of 100 families, but the need to increase the height of the development to achieve the size of homes in order to reach the target market niche of middle- and high-income earning professionals led to a final development of 75 flats and 16 detached houses (Figure 3.2). Having designed the development in such a manner as to be sustained by the available water resources, BCIL used the process of housing design to reduce the embodied energy in the development (Figure 3.3). The use of dry foundations, for example, was introduced as a way to save cement. Blocks of laterite, a traditional construction material, also reduced the use of energy-intensive concrete while contributing to the creation of a T-Zed brand of working with local materials and the 'earthy' look they create.

In addition to the novel design process, two forms of technological innovation were undertaken at T-Zed. First, the development of new context-specific technologies that sought to reduce energy consumption while maintaining the expectations of high-end consumers. For example, the hybrid LED and CFL system designed for street illumination was the result of an experimentation process that tested multiple illumination models. BCIL representatives highlighted how the air conditioning system and customized refrigerators were designed from scratch. According to BCIL this was needed in order 'to avoid the long-term risk of at least half the residents going for energy-guzzling ACs [air conditioners] or refrigerators that push up the Demand Load for projects of this kind for upmarket residential homes' (Interview 1, Developer, Bangalore, March 2010). Second, a range of existing technologies were innovatively applied within the context of T-Zed. Passive solar designs and natural ventilation were developed at the core of the project to reduce the heating and cooling needs of the building. Additional cooling was provided by simple landscaping solutions, house orientation and sky gardens. Solar water heaters were added to reduce the use of fossil-fuel-based energy resources. BCIL prided themselves on not doing technological but design innovation, emphasizing the idea of simplicity in their approach to housing: 'we use what is

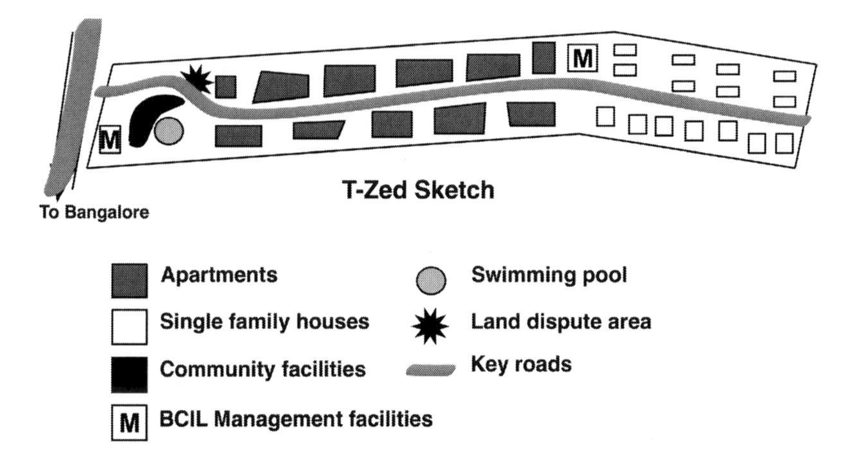

FIGURE 3.2 The layout of the T-Zed development. Sketch by V. Castán Broto.

FIGURE 3.3 A T-Zed block during construction. Photo © BCIL, used with permission.

known as "CKCS" software … common knowledge, common sense!' A local architect explained: 'in a sense, there are a lot of small design elements that are really innovative. But, by and large, most of the elements are taken from … what is already in the market, which is researched and used; we have not developed anything' (Interview 8, Consultant, Bangalore, March 2010). Yet nothing is taken for granted about T-Zed. From the gates of the compound to every room in each house, everything was thought of: the landscaping, the energy, water and waste systems, the construction materials and the decorations, the individual appliances and the habits of the residents, BCIL attempted to intervene at every level and in every dimension (Figures 3.4 and 3.5).

Social innovation

In seeking to intervene in the notion of an ordinary development to this degree, T-Zed relied not only on forms of technical design intervention, but also specific forms of social innovation. BCIL workers believed that T-Zed's success depended

FIGURE 3.4 A T-Zed single-family house, organic garden and LED lighting. Photo © BCIL, used with permission.

FIGURE 3.5 Collective facilities for building a community of eco-residents in T-Zed. Photo © BCIL, used with permission.

on responsible inhabitants: 'making people believe in these technologies is another challenge' (Interview 2, Developer, Bangalore, March 2010). The challenge of 'making people believe' was explicitly related to the innovative character of the compound: 'this is one of a kind, so it is kind of experiment … and customers are also not very sure; it is the first time they have been exposed, so convincing them is that difficult' (Interview 3, Developer, Bangalore, March 2010). To address this, T-Zed managers innovated by developing multiple strategies to control the behaviour of local residents, directly and indirectly. T-Zed attempted to 'coach' residents to equip them with strategies to live in the compound. This was a deliberate attempt to create low carbon subjects.

In terms of the direct forms of intervention, BCIL attempted to ban the installation of bathtubs in individual properties, as these were thought to create an unsustainable demand for water and disrupt the collective design of T-Zed. To extend mechanisms such as this, residents were invited to join BCIL in a collective project to develop community regulations for sustainable living: 'we were supposed to think of guidelines for things like what detergents to use, what cleaning things to use and ... what you can use in your gardens' (Interview 13, Resident, Bangalore, March 2010). BCIL, however, was perceived as being too directive and even unrealistic in some of their expectations, particularly around the consumption of food: 'a lot of people already have a tendency to look for organic vegetables ... but do you really want to mandate that?' (Interview 14, Resident, Bangalore, March 2010). In other cases, BCIL attempted to control the social use of resources 'by design'. For example, both the air-conditioning system and the refrigerator were developed to prevent the installation of unsustainable appliances once the house was sold. Automated control devices – called 'conscience meters' by BCIL – were installed to make residents conscious of both their use of energy and its environmental and economic costs. In these cases the design itself enabled a low carbon politics of life and residence in the compound, although its impacts were not always those predicted.

Overall, BCIL managers saw themselves as educators of the residents in T-Zed. In order for residents to interact correctly with the technologies, BCIL workers thought that residents should be made aware of how the technology worked and how it should be used. To this end, panels over the development explain and publicize its features. Posters act as a reminder of an implicit code of conduct inherent to those technologies, and thus they work as educative devices about how to live in the development. These forms of educational practices were developed through the whole life of the project. A programme called Zed-Interfaces provided spaces for meetings between BCIL personnel and customers:

> it was not like a regular ... sale process. So, it was not like the client came here, he was taken around a walk around the place, negotiated terms with ... the marketing team, spelled out his cheque, got under the payment loop ... closed! That [also] did happen but what happened after that was a series of meetings [Zed-Interfaces] with them, all of them together. Every month and a half, we used to meet every third fortnight at T-Zed.
>
> *(Interview 6, Consultant, Bangalore, March 2010)*

These were not only information-giving events, but acted as a means through which to exercise the desire to make residents understand their responsibility such that living in T-Zed required them to become low carbon subjects. In doing so, BCIL established an alternative model of the relationship between developer and client which only started, rather than finished, with the completion of the purchase.

Experiments in practice: making, maintaining and living T-Zed

From the process of design through to the forms of management through which BCIL sought to engage with residents, T-Zed constituted an example of both social and technical innovation for low carbon urban systems. Constituting urban responses to climate change in experimental terms requires the work of making, maintaining and living (Chapter 2). Here, we examine the ways in which these processes emerged and were engaged in the constitution of T-Zed, before turning in the final section of the chapter to consider their impacts and implications.

Making: from niche architects to green housing development

BCIL started as a loose grouping of individuals in the mid-1990s which brought together idealist architects and economists with visions of green housing as a potential business in Bangalore. The company began by building individual homes in the outskirts of Bangalore, in settings in which there was a lot of room for experimentation. The overall focus on innovation as the ethos of the company paid off, and in the mid-2000s BCIL saw its first opportunity for rapid growth. BCIL developed a niche for its green building concept by building second homes in low-density developments. This provided the opportunity to both refine the product that BCIL was selling and its own institutional identity as a hybrid between private developer and advocacy organization. These pilot projects were crucial to developing not just a green housing product but a grammar that would enable BCIL both to imagine and market new solutions for green housing. Finding means to design and calculate the low carbon benefits of this new form of housing was a crucial step in this process. T-Zed was the first attempt to move beyond the scope of individual houses and expand into a larger scale. BCIL purchased a plot of farmland in the area of Whitefield at the edge of the city (Figure 3.1). The location was selected to make green housing available for middle- to upper-class families working in the IT and outsourcing sectors. T-Zed was thus built upon broader structural changes in Bangalore. To describe Bangalore as the Silicon Valley of India is now considered a cliché, but it neatly captures the clustering of IT industries since the late 1970s and the aspirations of some city managers. Bangalore as a hub for international investment and capital is a vision that reflects only a partial view of the economy of the city, but one that has had a powerful role in shaping urbanization, particularly through the growth of the middle classes who work in these sectors from whom there is a distinct demand for Western-inspired products, services and housing, as well as a new articulation of global discourses of environmental protection.

Forging alignment of economic, environmental and urban development interests in the notion of low carbon housing was BCIL's key objective in T-Zed. In keeping with the ethos of innovation that they had established, the assemblage of T-Zed as a housing development was critically dependent on a continued process of experimentation in which all possible interventions that could be done had to be tried. These interventions followed a process of trial and error which enabled innovation to emerge as the project unfolded, for example in the testing of new forms of LED lighting, the development of bespoke refrigeration systems, and new forms of air conditioning. One of the more successful elements of this process of innovation was that of the LED lights which not only demonstrated the feasibility for street lighting but also supported the further development of the lighting company Flexitron. In T-Zed, Flexitron underwent a painful process of technical innovation: '... we kept making and breaking and making and breaking ... maybe about 500 models ... I blew up a lot of LEDs! It's a process where ... we need to put in all our best efforts to see how the technology works!' (Interview 6, Consultant, Bangalore, March 2010). The breakthrough came when the LED lighting systems were adapted to the T-Zed requirements

> we brought in a new kind of LED ... like a sodium-vapour light ... it mimics the [conventional] lamp in the same way of light ... So that we standardized and started using for the street lighting! Now, that has become a standard in T-Zed!
>
> *(Interview 6, Consultant, Bangalore, March 2010)*

Eventually, not only did Flexitron find a solution for the specific lighting problem in T-Zed, but they were also able to 'design and develop an emerging technology at T-Zed, which is LED ... to kind of demonstrate that [it is a] much more efficient lighting system' (Interview 6, Consultant, Bangalore, March 2010).

While this process of innovation by trial and error led to some new technologies being adopted, it also resulted in considerable delays in the completion of the development. Even when the technology was well-known, such as with the traditional practices of constructing with compressed blocks, a lack of skills among the workforce (accustomed to ready-made materials used in conventional developments) made it difficult to deliver the development in the projected time. Local materials, such as laterite, were difficult to cut. Initially, careful experimentation led to the building of beautifully polished laterite walls in the communal areas of the development, but as delays accumulated and influenced the relationship between BCIL and its customers, the building process had to be accelerated. Less careful cuts resulted in rough laterite blocks, which often had to be covered with cement or aided with additional materials to install windows and other features. At other times, problems emerged with the delivery of specific materials, demonstrating a certain resistance among providers to shift conventional practices:

> the guy who is actually doing it ... [would say:] 'hey, hey, hey ... don't even tell me about this carbon thing ... I don't know what this is about! I need

> to make my business succeed!' … which means he is not actually willing to invest in the innovation that it takes to bring this figure [the price] down!
>
> *(Interview 4, Developer, Bangalore, March 2010)*

Technical difficulties and a concern with lowering standards halted the making of the project. Much of the weight was borne by BCIL workers and architects, which in turn led to lack of continuity in BCIL's workforce. According to one architect, few people in BCIL work in the company for more than two years. Another BCIL member explains:

> A lot has happened, we've had 10 years, we have lots of [fits] and starts, we had lots of people not having the energy to run with us … it's not easy to run with this! … You need the energy … you need the conviction! That is the only word! … [My only regret is] that I have not been able to carry enough people with me for enough time … 1995 to today, there are only two more people after me in this organization who have come with me from then!
>
> *(Interview 4, Developer, Bangalore, March 2010)*

This emotional investment in the ideal of T-Zed and what it could deliver was both essential to aligning the development as low carbon, economic and socially desirable, but was also regarded by some residents as leading to too much confidence in untested technologies: 'maybe they could have focused on a few larger impact things and gotten that right [instead of implementing 62 different technologies] … there was too much confidence [but] maybe you needed a little caution over there' (Interview 14, Resident, Bangalore, March 2010). To counter such views, the forms of engagement with residents that BCIL undertook, while originally intended to be orientated towards educating customers in how to use the development, also served to make customers aware of the costs of change personnel, learning problems, changes of people and eventually share the anxieties and risks in the delivery of T-Zed. Enrolling future residents was a strategy to deal with the difficulties of rendering the experiment technical, manageable and calculable.

Thus, making T-Zed involved calculating, codifying and realizing particular forms of innovation and their contribution to addressing the problematic of achieving zero carbon housing while simultaneously building a compelling emotional case that would draw the clients to assemble the resources necessary to make T-Zed possible. Thus, forging alignment and rendering the experiment technical was not sufficient: the enrolment of key actors, especially residents, was a central factor for the success of the project. In doing so, BCIL not only had to assemble a new range of materials and technologies through which the development could be designed, but also through their dialogues with residents and professionals, they had to redefine client-customer relations. This also led to hybrid portrayals of residents as environmental activists or BCIL as a committed civil society organization (rather than developer) motivated by higher aims rather than by profit. The extent to which these ideals corresponded to the actual performance of roles during the project is questionable – as BCIL focused on long-term profits and residents emphasized their right to maintain their lifestyles – but it is clear that the construction of an ambiguous relationship was necessary to draw together the different actors involved in T-Zed. As we will see in the following sections, this was a bumpy process that led to several stages of negotiation, contestation and the reconfiguration of some aspects of the experiment.

Maintaining: embedding low carbon development

BCIL have continued to intervene in the maintenance of T-Zed, arguing that, due to the exceptionality of T-Zed, they are required to provide customized care for both the residents and for the development itself. Because of the two-year delay in the delivery of the project, partially arising from the experimental process discussed above, some residents moved in when work was still going on: '[We moved in] about 4 years ago, 2006, it was half finished: I was the first one to move, but we were so desperate that we had to move. It was very difficult with all the dust, noise, people all around, but it is good now' (Interview 10, Resident, Bangalore, March 2010). Some of the houses were completed even after residents moved in: 'after I moved they constructed the bathroom, they had scaffolding all over it, and I had small children' (Interview 12, Resident, Bangalore, March 2010). 'Construction was still going on still two month back' explained one resident in March 2010 (Interview 11, Resident, Bangalore, March 2010). Another explained the consequences of this for her family:

> at some point it just got to me that you are staying here and you are continuously hearing the sound of somebody cutting stone … because they are replacing something or they are changing something, right! And, the kids goes down to play … because things are not finished … there are like levels and there are rough edges where you can hurt yourself … it seemed to me: 'okay, you are trying to do all the environmental stuff but you are not really taking care of the people who are here'. Because, the kids can't play freely … because this is a construction site … to me the whole point is lost!
>
> *(Interview 13, Resident, Bangalore, March 2010)*

For the developers, maintenance issues in T-Zed are not so different from those experienced in other developments:

> Initially we had dampness. Initially!!! Because of the rain or so. And also, the walls, the customers they have painted in their own way and then we have found bubbles in the paint, because of the paint has not been absorbed properly; then plumbing, then power problems, but normal maintenance issues, there is not anything specific for T-Zed.
>
> *(Interview 3, Developer, Bangalore, March 2010)*

However, many of the maintenance issues encountered, including the 'bubbles in the paint', relate to specific materials and technologies used only in T-Zed. These innovations had additional maintenance costs, which surfaced only in the advanced stages of the development, and have led to BCIL becoming not only housing developers but custodians of the T-Zed compound. It appears that BCIL may not have taken into consideration the difficulties of maintaining such a development from the outset:

> over time, they realized that we have these things that had not necessarily worked out! So ... they are changing, for example, the system ... the water purification system, right, so now the same purification now works for the water coming in all the taps, right. Earlier, they were supposed to have different taps for water that was coming in to drink as opposed to that which was coming in the bathroom and stuff like that ... once they realized that some of these lifestyle things that they were trying to advocate were not going to necessarily work out, they just went there and changed the system.
> *(Interview 13, Resident, Bangalore, March 2010)*

As this resident suggests, some of the technologies that were central to the making of T-Zed as an experiment have simply not worked. Technical problems in implementing an innovative refrigeration system were coupled with residents' perceptions of the fridges as bulky, over-designed and generally not fitting with the overall spirit of the place, despite their low carbon credentials. In the process of managing residents' expectations, BCIL allowed them, after an initial ban, to install their own appliances. These examples show that at the same time as engaging in the material repair and upkeep of the new forms of innovation present at T-Zed, the development has also required investment in terms of engagement with residents to enrol them in the operation of its innovative technologies, processes which are not always successful.

T-Zed has also required forms of metabolic adjustment to enable its insertion within and maintenance through the material flows of the city, particularly in terms of the ways in which T-Zed has intervened in the material and ideological circulations of water. T-Zed was originally intended to operate independently from the municipal water supply, with a strong discourse of creating water and energy self-sufficient communities in which 'you are not taking any electricity [nor] water from the state ... you generate your own [energy and] water and then use it' (Interview 9, Consultant, Bangalore, March 2010). Yet achieving self-sufficiency in practice has been challenging. As outlined above, the design of the development and its rainwater harvesting system was shaped by calculations of the water resource use that would be required. However, a period of successive droughts has meant that the rainwater harvesting system has never operated as envisaged. Pressed by the residents' demands, BCIL resorted to the dominant strategy for most gated community developments in the area: to dig boreholes. In offering a high quality resource to its residents, T-Zed redirects circulations within the urban water system in ways very different to those intended by the notion of self-sufficiency, serving to diminish the very system that it set out to protect. Simultaneously, poorer residents in the immediate nearby areas have found ways to collect water from T-Zed, through informal means or paying bribes to gatekeepers. This is seen as a threat by residents to the newly-secured water resources on the development, who do not recognize the irony of how their own practices are encroaching on the collective resources of the area and contributing to limiting resource availability for the wider urban population.

The processes and politics of maintaining T-Zed demonstrates the ways in which implementing innovation requires progressive adjustments to fit technologies to local contexts, and the means through which this in turn creates new forms of circulation – of resources, technologies, ideas – through the development and the city. In seeking to sustain the experiment and its innovative qualities, BCIL sought to continually engage in its redesign and re-engagement in order to ensure that its credentials were maintained and reinforced. This distinguishes T-Zed from other conventional housing developments that are delivered as finished and where the developers disappear from the scene. In this sense, T-Zed is an ongoing living project. Simultaneously, however, maintaining the experiment has also required adjusting it to both the material conditions of its implementation and the residents' expectations, in a process that has drawn from conventional construction and consumer practices more than BCIL or residents openly admit. This has resulted in a series of micro-practices of maintenance and repair that construct the appearance of normality in what is a very unconventional development. The notion of self-sufficiency, advanced here under a carbon reduction flag, has served to contribute to increasing demands on the metabolic circulations of the city and to the securitization of resources for its future viability. While T-Zed provides a means through which new forms of low carbon innovation have been tried, redesigned and brought into housing developments, the insertion of the development in the city has reinforced new urban metabolisms which serve to direct resources to the urban elites and sustain resource scarcities, particularly in terms of the lack of water and increasing price of electricity, that are affecting the whole city.

Living: installing and resisting low carbon innovation

> We want to be like this, we want to save the trees, we want to teach our children, we want our children to know that [this] is how our quality of life has improved.
>
> *(Interview 10, Resident, Bangalore, March 2010)*

Through T-Zed and subsequent projects, BCIL sought to create not only a development but also the ideal resident that would enjoy and contribute to the growth of the notion of zero carbon housing through enacting their moral duty to reduce carbon emissions. As outlined above, T-Zed combined various forms of technical and social innovation designed to act both directly and indirectly towards engendering new forms of low carbon subjectivity – from the provision of energy and water, through conscience meters to the encouragement of particular practices concerning bathing, eating and mobility. Through these interventions and innovations, customers are represented not only as private property owners, but as recognizing their wider responsibilities:

> [residents] are all eco-friendly people … who have it in their heart that they need to do something for this community; and there are people from all backgrounds, the cream of the society. From software to doctors to people in the retiring ages … there is a community as a whole … They figure they will benefit from it … not only them but the generations ahead of them as well. It's just not the persons who are living here, its generations of people who are going to live in, their kids, their grandchildren. … [This] makes an impact on the country and the surrounding world.
>
> *(Interview 1, Developer, Bangalore, March 2010)*

This moral duty is not one that entails sacrifice, but rather one that can be bought without compromising the lifestyles of the accommodated classes. The role of the developer is, thus, to provide the solutions that will simultaneously meet the demands of lifestyle and the moral duty of customers. However, to help customers to develop a green conscience, the developer also needs to deliver the right product:

> He is buying because the location is right, the finish is right, the cost is right and he has the confidence in us. … it is not the Green factor! He is not buying because it is sustainable! … [but for] REEFF – Reliability Economy Efficiency Function and Finish. IF I don't have that, I don't have my customer with me. … This is, of course, a function it has to serve … but once the client has bought it, we must be excellent of finishes or then they are not interested in buying!
>
> *(Interview 4, Developer, Bangalore, March 2010)*

Thus, T-Zed, and similar developments that are emerging in Bangalore in the wake of its success, are presented as convenient products that meet both customers' lifestyle requirements and existing moral concerns. Residents are exhorted to regulate their own behaviour, but this is done under the watchful guidance of BCIL, whose personnel think of themselves as educators. Subjects are produced not just by those who are expected to behave in a particular way, conducting their own conduct, but also by those who place such expectations on their behaviour.

Much of this process has led to the production of knowledge about what constitutes appropriate low carbon behaviour. Through their work at T-Zed, BCIL have developed concrete guidance to lead low carbon lifestyles. Even though residents may have been attracted to T-Zed because of their personal green commitments, some of them explain that living in T-Zed has influenced their views: 'maybe our thinking has changed because before living in T-Zed we never thought about waste segregation, CFL bulbs, centralized [energy] … after we moved here we started to think environmentally' (Interview 10, Resident, Bangalore, March 2010). Residents' comments suggest a practical element embedded in the development of environmental citizenship because 'only when you do [everyday actions do] you realize how important the environment is' (Interview 10, Resident, Bangalore, March 2010). These practices have extended beyond T-Zed, influencing other choices:

> I am more conscious of my carbon footprint. And about … not using plastics; I never take plastics when I go shopping, never accept plastic bags. And being a bit more conscious, using public transportation … I see people not doing it as a big shame.
>
> *(Interview 12, Resident, Bangalore, March 2010)*

Overall, within T-Zed the creation of carbon subjects has been successful to the extent to which residents accept that they have a green conscience and such conscience moves them to act in particular ways. Indeed, considerable effort goes into visible symbolic acts which ratify the green status of residents, further cementing the low carbon status of T-Zed:

> [residents] do a lot of awareness programmes on eco-friendly living and all these things. They participate in walk-a-thons, marathons, all these things … there are all these other things that are being celebrated like World Wide Fund for Nature, all these things. At nights the lights are switched off … all those things.
>
> *(Interview 1, Developer, Bangalore, March 2010)*

However, the creation of carbon subjects has also served to test the limits of the experiment, through the encounter between everyday practices and the ingrained assumptions of the development, as well as through residents' contestation of what it means to be a low carbon subject. The use of the development in everyday life has unveiled some of the inadequacies of design and the mismatch between design and residents' expectations:

> in today's floors, we almost want to look at the floor and brush our teeth or adjust our hair! We like that extra, mirror … you know, kind of appearance, okay. But, those [natural] floors were never meant to be like that! You can polish them but they will become dull.
>
> *(Interview 6, Consultant, Bangalore, March 2010)*

The reverse is also true, where expectations on the part of BCIL about how it might be possible to live in T-Zed have come to be contested. For example, some local residents followed BCIL advice of growing organic vegetables, but after some time they thought that it was unviable and required too much personal work considering their personal priorities. A similar issue emerged regarding the installation of bathtubs. As outlined above, BCIL banned bathtubs on the grounds that they use too much water in comparison to showering. However, residents perceived this as an attack on their own life choices and chose to install bathtubs regardless of BCIL recommendations. Levelling such a critique at the bathtub, while simultaneously having a swimming pool and an air conditioning system as part of the development, puzzled those residents who saw taking a bath as part of their lifestyles. Having already put themselves forward as morally engaged by buying into the zero-carbon ideal, some local residents perceived that additional requirements to conduct their lives in this manner were unrealistic and unfair. Furthermore, other residents found living with novel technology a burden, in the sense that they did not want to be continuously confronted with the technologies they dwell with and whether they work or not. During the construction, for example, improvised solutions had to be found to fit around novel materials which did not behave exactly as designers had predicted. Customized fridges did not work properly: not only they did not cool the food adequately, but they had been designed with limited space inside. As an important symbol of mass-culture, some residents decided to buy their own fridges, adjusted to their own needs. Similarly, technical problems with the air conditioning system and the smart metering remain. In this context, some residents argued that the process of experimentation could have been eased by not going to the extreme for every aspect of carbon control from the construction process down to regulating the lives of individuals living in T-Zed. Yet it is precisely these forms of excessive innovation that have been central to the identity of T-Zed as an experiment, and its goal of positioning itself as a space of demonstration and learning.

Indeed, central to the momentum of T-Zed has been its ability not only to engage residents in the project of zero carbon, but to enrol a wider network of actors and interests in drawing on the principles and ideas of T-Zed and putting them to work elsewhere. The idea is that the emergence of green consumers in T-Zed can be replicated elsewhere:

> small changes begin from home, it's true. People who come see these things; they learn a lot of things from here like maybe the guests that are often here. All these people, they also learn some values from here. They go back home and try to implement small things what they learn from here. Yes, definitely it is helping in one other person spreading the awareness.
>
> *(Interview 1, Developer, Bangalore, March 2010)*

So, by being close to T-Zed other people 'can watch and learn' (Interview 5, Consultant, Bangalore, March 2010). Even residents explain their own experience of spreading awareness:

> when you say T-Zed there is a certain awareness for it and people ask you what it is. And then you talk about it and it gives the idea to a lot of people

> that it sounds nice and a lot of people ... think 'oh, if we had more options like this, we would do it!' ... So I think that one of the things is that this is a proof that all this stuff can happen so people at least think about these things!
>
> *(Interview 14, Resident, Bangalore, March 2010)*

Beyond these informal encounters, residents have sought explicitly to liaise with neighbouring gated communities and move them to adopt what they regard as greener lifestyles. They have started by trying to minimize the use of pesticides, because they perceived it affected them directly:

> the neighbouring complex, they really get the mosquitoes in the evening in the mosquito season: so it's like fumes of DDT coming into our campus and settling on the soil and everything. So, whether I use some pesticide or not, I won't really make that much difference ... And we've been going there and talking to them about stopping fumigation. We even gave them an article about DDT and what it does. ... They stopped for a year! They did that, they did stop and I think people were affected by that article; but they started again but this time the fumes are not so strong, it doesn't come right into my house.
>
> *(Interview 13, Resident, Bangalore, March 2010)*

Waste segregation, the great success at T-Zed, may constitute a point of cooperation between residents and neighbouring compounds: 'I think people have been talking about using our bio-gas plant ... [instead of] sending all their vegetable waste into the municipal garbage' (Interview 13, Resident, Bangalore, March 2010). This may be early days, but there is a clear intention to reproduce the models of green living enacted in T-Zed on behalf of both the developer and the residents, contributing to ensuring that the innovation embedded within T-Zed is continually re-enacted and made real.

Impacts and implications

In the wider context of urban development in Bangalore, BCIL has positioned itself as a hybrid public–private body which, though profit motivated, adopts regulatory functions by creating new standards for the design of housing developments together with specific behaviour requirements from T-Zed residents. This form of experimentation is often explained by the proponents of T-Zed as emerging from a sense of professional responsibility to respond to global challenges. In this context, India's position towards climate change action does not cohere with the ethical and behavioural demands of some of the actors operating at the city level in Bangalore. As has been seen elsewhere internationally, inaction at the national level does not lead to the paralysis of other sectors of society, but rather fosters alternative forms of action and social innovation, such as in the case of BCIL which emerged 'out of a sense of anger' (BCIL n.d.) towards the existing status quo in the housing sector in Bangalore. Whilst clearly positioned in response to a set of emerging concerns about the responsibilities of middle class Indian citizens in relation to climate change, T-Zed also needs to be read in relation to the dynamics of land, resources and consumption in Bangalore. As a private and gated housing development, the experiment served as a means through which to test which ideas and technologies might be successful in the housing market of Bangalore – forging alignments and rendering technical the problem of green housing – and as a space through which to develop new forms of urban subject – green housing developers and green housing consumers – which would form the basis for new forms of urban economy. The constitution of low carbon subjects as green consumers has been crucial for the success of the experiment.

In fostering the notion of the viability of a green housing sector in Bangalore, four aspects of T-Zed have been particularly significant. First, the development served as the basis for experimentation and innovation with a whole host of new technologies. As discussed above, many such technologies failed to take hold at T-Zed and have since been abandoned. Others were embraced, such as the LED technologies developed by Flexitron which have since moved beyond the development to be taken up across India. Second, T-Zed provided a means through which to challenge existing preconceptions about the cost of green housing. For BCIL, T-Zed succeeded in debunking the idea that green houses would cost about 10 per cent more than conventional homes. BCIL, however, finds that innovation has come at a cost in terms of 'the time lost in planning, zeroing down on technologies, innovating, and getting eco friendly materials all of which are an opportunity cost, therefore a compromise on profits'. Third, it enabled BCIL to establish themselves as a leading brand for low carbon housing, opening the way to future innovation: 'it was important for BCIL to ... establish ... a strong sustainable building methodology into the future ... and create ... an already existing precedent for itself into the future' (Interview 6, Consultant, Bangalore, March 2010). Finally, T-Zed provided the means through which to foster the 'green housing consumer' as a subject that could be enrolled in new forms of housing market in the city, establishing the appetite amongst the wealthier residents of Bangalore to engage in the discourse and practice of zero-carbon in their housing decisions and everyday practices.

Overall, the transformation of green housing from a bespoke to a commercializable product has arisen from the interests of the private sector that now sees a double dividend in improving their image and securing resources for the growing middle class in Bangalore. The role of T-Zed in privileging some solutions for carbon reduction over others (for example by promoting larger houses and a communal swimming pool but controlling the use of water in-house) contributes to stabilizing the meaning of what counts as zero carbon in the context of Bangalore. This is reproduced in new eco-city initiatives and less ambitious gated developments that also build on the knowledge developed and gathered through T-Zed and subsequent projects by BCIL. BCIL's new projects have not involved the level of ambition, innovation and risk seen at T-Zed, reflecting the challenges of dealing with the processes of maintaining and living discussed above, but have instead seen new forms of 'experiment' with more conventional materials and technologies that seek to maintain BCIL's reputation as at the fore of developing sustainable housing whilst also sustaining the appeal of climate change interventions in the housing market as a consumer product. In this sense, the experience of T-Zed has served to mainstream climate change within the emerging green housing market in Bangalore; on the one hand ensuring issues of efficiency and resource security are increasingly being taken into account in the housing being developed in the city, but at the same time reducing the potential for achieving zero-carbon development. In so doing, BCIL themselves have moved ever closer to the role of a mainstream developer, concerned with issues of efficiency and delivery for customers. At the same time, Zed Life, described as a BCIL company, seeks to develop the green housing niche in Bangalore through the sale of new housing developments, eco-friendly technologies (air conditioning, solar energy systems) and furniture.

Beyond the direct impacts on the emerging new market for green housing and associated technologies and products that T-Zed has helped to sustain, T-Zed and its subsequent development also has wider implications for the city. At the heart of the zero-carbon project of T-Zed was its ambition to be self-sufficient in energy and water resources. The processes of innovation, both those focused on the development of specific technologies – for lighting, cooling, heating, washing and so on – and those which sought to attend to the everyday practices of residents were developed with these principles in mind. Yet, while laudable in intent, in the context of urban development in Bangalore, the implications of the intention to be self-sufficient are more ambiguous. On the one hand, the notion of autonomous provision means that the developers and residents of these compounds have limited incentive to participate in the universal provision of urban services, can opt out of existing infrastructure, in turn potentially contributing to a process of splintering urbanism (Graham and Marvin 2001). On the other hand, the actual practice of autonomous provision has proven hard to realize, such that resource securitization is only achieved through reconnecting the development to the metabolic flows of water and energy through the city, encroaching on the collective and future resources of the city and suggesting that the promotion of this form of development as sustainable (and equitable) is highly equivocal. This is particularly important for while in T-Zed sustainability was often regarded as measurable through processes of carbon accounting, beyond the technocratic discourses of developers, NGOs and consultants emphasized the links between sustainability and justice in the context of urban India such that: '... in India it's not about climate change at all ... it's all about the social justice part of it ... and both of them have to go together' (Interview 7, Consultant, Bangalore, March 2010). Attempts at advancing climate change objectives at the expense of social justice is likely to raise conflicts, particularly when they are, for the most part, absent from the rationalization and motivation narratives of green developers and thus question the kind of climate change interventions advanced in projects like T-Zed. Ultimately, T-Zed's model does not challenge the fundamental inequalities embedded in the constitution and expansion of a city like Bangalore (see Nair 2006). Instead, T-Zed has played a mediating role to adapt some features of the existing construction industry to circulating discourses of sustainability and climate change. This is vital and important in the context of rapid urbanization and growing levels of middle class consumption, but at the same time fails to address the challenges of social and economic justice that are at the heart of the politics of both urbanization and climate change.

References

BCIL (n.d.) *BCIL: An overview*. Bangalore: B. C. I.

Government of Karnataka (2007) *Bangalore Master Plan 2015 – Vision Document*. Bangalore: Government of Karnataka.

Graham, S. and Marvin, S. (2001) *Splintering Urbanism: networked infrastructures, technological mobilities and the urban condition*. London: Routledge.

Mukhopadhyay, P. and Revi, A. (2009) 'Keeping India's Economic Engine Going: Climate Change and the Urbanization Question'. *EPW*, XLIV(3): 59–70.

Nair, J. (2006) *The Promise of the Metropolis: Bangalore's twentieth century*. New York: Oxford University Press.

PricewaterhouseCoopers (2009) *Global City GDP Rankings 2008–2025*. London: PricewaterhouseCoopers.

Ranganathan, M., Kamath, L. and Baindur, V. (2009) 'Piped Water Supply to Greater Bangalore: Putting the Cart Before the Horse?'. *EPW*, XLIV(33): 53–62.

4

SOCIAL HOUSING AND CLIMATE CHANGE

A sustainable housing project in Monterrey, Mexico

Introduction

Mexico is regarded as a leader in climate policy at the local level, particularly after the pioneering example of the Climate Change Action Plan for Mexico City and the subsequent efforts in promoting climate change action in cities through the Conference of Mayors in Cancun in 2010. This focus on urban responses has been materialized in concrete policies at the local level, but it has also been a crucial part of national policy. As Mexico has sought to both reinforce its leadership commitments and secure international climate change resources, it has turned to revise key national policies in its search for opportunities to implement climate change interventions. Without challenging its current reliance on fossil fuels and other structural issues that serve to produce significant levels of GHG emissions, there has been a concerted effort by the national government to green one of its key policies: the provision of housing for low-income citizens. The resulting Green Mortgage programme received the 2012 World Habitat Award as an example of how to address climate change at the same time as social concerns.

This chapter examines an experiment that supported the initial development of the Green Mortgage programme: Vivienda de Diseño Ambiental (ViDA; literally, 'environmentally-designed housing'). Implemented in the town of Escobedo, within the urban conurbation of Monterrey, ViDA has enabled the problematization of housing provision and the need to ensure access to affordable energy resources for residents over the long term by bringing diverse but complementary interests into alignment. However, as the project became inhabited and integrated into the social and economic circulations of Monterrey, it was not successfully maintained or lived. As a result, the development has merged with its landscape and its experimental characteristics have lost visibility. It is perhaps paradoxical that ViDA's loss of visibility and its normalization within the construction industry has allowed for the reproduction of its essential elements at the national scale through the Green Mortgage programme. This shows that the art of governing involves not only processes that problematize and make calculation visible, but also those which serve to normalize the process of calculation.

Urban development and climate change in Monterrey

Monterrey is the rapidly growing capital of the Northeastern State of Nuevo León. Its proximity to the USA–Mexico border led to the establishment in 1900 of The Fundidora, a steel foundry company that has shaped the history and development of the city. Although The Fundidora went bankrupt in the 1980s, Monterrey was by then established as the industrial capital of Mexico, attracting capital from within the country and across the border. The proximity to the USA allowed Monterrey-based companies to target lucrative American markets while accessing an extensive pool of cheap Mexican labour. Today, Monterrey hosts an array of national and international companies, especially in the peri-urban areas of the city, which have led to a polycentric model of development in which urban growth closely follows the sources of employment. As a result, Monterrey is part of a conurbation which includes nine cities (each of which is self-governing): Monterrey, San Pedro Garza García, Santa Catarina, Guadalupe, San Nicolás de los Garza, Apodaca, General Escobedo and Juárez Monterrey, Escobedo, Apodaca and Guadalupe. While the last national census of Monterrey city (2010) estimates its population as 1.1 million inhabitants, the continuously populated metropolitan area may host up to 3.7 million inhabitants, 88 per cent of the inhabitants of the whole state of Nuevo León (Nuevo León Government 2012). Although both industrial and population growth have declined since the 1980s, Monterrey continues to attract migrants from Mexico and other countries in Central America who are looking for employment opportunities. Levels of social and economic deprivation, however, are also high. A study of marginality indicators in the Monterrey metropolitan area estimated that more than a quarter of the population lived in households supported by less than two minimum salaries, and many households lacked access to basic resources such as a non-precarious ceiling (15.2 per cent), drainage (18 per cent) or fresh water (12.4 per cent) (Montes and Ortega 2003).

The proximity of Monterrey to the Mexico–USA border also promotes the circulation of technological development ideas and knowledge associated with an industrial culture that has promoted the establishment of Monterrey as a centre for innovation. As the owner of a sustainable business explained:

> Being in Monterrey is a privilege because of its proximity to the US market … the availability of the best Mexican Universities and a cheap and accessible labor force. The local culture is attached to industrial development and innovation, … a tradition … that we have grown with.
>
> *(Interview 1, Corporation, Monterrey, May 2010)*

Institutions such as the Monterrey Institute of Technology and Higher Studies (the 'Tec de Monterrey') and the Universidad Regiomontana (both of which receive additional funding from locally-based corporations) have provided support for an innovative business sector, marked by an ambition of reaching the neighbouring markets in the US. Since 2008, however, the proximity to the Mexico–USA border has been influential in less positive ways, as it has led to an increasing presence of drug cartels, which brought with them a culture of violence that has spread throughout the city. In addition to this growth in violence, drug trafficking has also been associated with a new set of circulations, including the abuse of migrant and vulnerable populations, weapons trafficking, corruption and the securitization of public spaces, all of which are shaping the city in new ways.

Against this background of industrialization, urbanization, growing economic development and social challenges, Mexico has also embraced issues of sustainability on the international stage. In particular – and reflecting the changing dynamics of the international climate change policy regime in which growing pressure has mounted on rapidly industrializing countries to play a role in efforts to reduce GHG emissions – Mexico has sought to show leadership in this area. In July 2013, Mexico launched the National Climate Change Strategy, which will guide the country's climate change policies for the next 40 years, replacing the previous 2007 Strategy. The new strategy focuses on the transversal incorporation of climate change to other policies and the development of financial instruments to advance climate objectives. The financing of carbon reduction measures alongside social housing thus resonates with the government's overall approach to climate governance (Federal Government of Mexico 2013). The incorporation of climate change into local policies, best exemplified by the development of the Climate Change Action Plan in Mexico City, was undertaken by building upon previous experiences of dealing with pollution by grouping ongoing or planned activities that could address climate mitigation, but without directly tackling urban development as an area where emissions reductions can be achieved (Interview 2, Local Government, Mexico City, May 2010). Overall, the strong governmental emphasis on emissions reductions through inventories, energy efficiency and renewable technologies such as biogas contrasts with the continued, if declining, economic importance of fossil fuels in the economy.

In Nuevo León, the Climate Change Action Program 2010–2015 emphasizes climate change mitigation with an objective of reducing 1,558 million of tonnes of CO_2 equivalent with an investment of 14 million pesos for 20 actions in five sectors: transport; energy efficiency in houses and SMEs; nature conservation; and waste management. Most measures, all of which will have a direct impact in Monterrey, promote an individual-based approach to reducing emissions and adopting climate change responsibilities, rather than being integrated into broader structural economic changes or into the process of urban planning. Rather than offering a coherent and integrated agenda, the Climate Change Action Program here, as elsewhere, emerges in the context of experimentation in which ideas about what climate change action might involve have been heavily influenced by existing and ongoing projects, which serve as examples of what is possible and, in doing so, render climate change as a problem amenable to solutions.

Urban planning and the provision of social housing

In the context of rapid economic development and high levels of inequality, the provision of housing is a critical political agenda and one that serves to shape urban development directly, with planning often confined to the provision of services in response to the development of new residential areas of the city. For working class residents in Monterrey, two types of housing predominate – popular housing and social interest housing. Popular housing relies on incremental self-build strategies to establish single-family houses in urban and industrial peripheries in neighbourhoods called 'colonies'. Although this type of housing often relies on informal strategies of land acquisition and construction, many colonies have achieved formal recognition after their establishment. In contrast, social interest housing is driven by Infonavit (Institute of the National Fund for Housing for Workers), a financial institution operated as a partnership between the government, business and workers. Infonavit collects a percentage of salaries from workers and, after a saving period, provides them with mortgages to access the market for social housing. This system has helped to establish a continuous demand for social housing, provided through highly standardized and homogenous housing developments in accordance with standards defined by Infonavit, most often in the suburban areas of Mexican cities.

The regularity and uniformity with which social housing is produced led one government official to describe Infonavit as 'a hen that lays homes' (Interview 3, State Government, Monterrey, May 2010). For Infonavit, facing continual demand for new housing, the overall focus is on the quantity of housing provided rather than its quality or the potential profits that may be realized. In 2011 alone, Infonavit supported the construction of more than half a million houses, which they estimate to house 1.7 million people (Infonavit 2012). In so doing, Infonavit also supports the for-profit construction industry in Mexico, which is a key partner in this institution. The reach of Infonavit is significant not only for the housing market but also for the Mexican economy at large. In addition, CONAVI (the Federal Housing Institute of Mexico) and organizations such as the Housing Institute of Nuevo León support housing developments for informal workers who have no access to Infonavit at even lower prices with the help of subsidies. In both cases, the model of urban development is very similar: it relies on building on cheaper land in peri-urban areas in the city using highly standardized templates for single-family houses, often built at the minimum possible cost. Table 4.1 provides a comparison of the two housing types associated with housing strategies for low-income citizens in Mexican cities.

Despite the success of the Infonavit model, popular housing continues to be a recurrent strategy not only for those citizens who cannot access Infonavit credits but also because it provides greater flexibility and the capacity to adapt to multiple

TABLE 4.1 Comparison of social interest housing (formal) and popular housing (informal) in Mexico

Social interest housing (Infonavit)	*Popular housing (self-built)*
• Smaller houses	• Bigger houses
• Bigger costs	• Lower costs – incremental construction
• No personalisation	• Tailored constructions
• Standardised	• Lack of standards
• Technical and architectural innovation	• Low quality materials and processes
• Land security	• Land may not be secured
• Standardised infrastructure requirements	• Problems with infrastructure access
• Delivery of finished product	• Housing remains 'in the making'

livelihood strategies which emerge in the Mexican city. However, such flexibility comes at cost both in terms of the quality of the construction and access to infrastructure. While the popular housing model relies on the private initiative of would-be residents, the social interest model enrols multiple institutional levels in the process of making the city. By making available workers' capital to the construction industry and facilitating its economic growth, the government plays a key role in housing provision. Infonavit is at the centre of this alliance, as the representative of the Housing Institute of Nuevo León explained:

> Infonavit is the financier: a developer that makes their homes tries to sell them to people who have the right to request a credit to Infonavit ... The client is Infonavit, and because Infonavit is the boss they determine the rules.
> *(Interview 4, State Government, Monterrey, May 2010)*

As we explore below, by seeking to innovate within the existing and dominant model of the provision of social interest housing, ViDA created an aperture through which new ways of determining the rules of housing design and provision could be developed that matched the need to provide housing at the scale and price required to satisfy popular concerns and to maintain existing structures and processes of urban development in Mexico, whilst at the same time bringing issues of climate change and sustainability into this mainstream agenda.

Vivienda de Diseño Ambiental: experimenting with housing for climate change

Vivienda de Diseño Ambiental (ViDA) is a social-interest housing project completed in 2007 in the town of Escobedo, Monterrey (Figure 4.1). The project comprises 58 single-family houses for the families of low-income salaried workers. Each house, sold at around US$20,000, comprises 64m^2 plus public and private outdoor spaces. Driven by the ambition of reducing the costs of living for such families, as well as mitigating climate change, the development is designed to take advantage of passive heating, cooling and ventilation, and to incorporate various sustainable technologies (Figure 4.2). Its

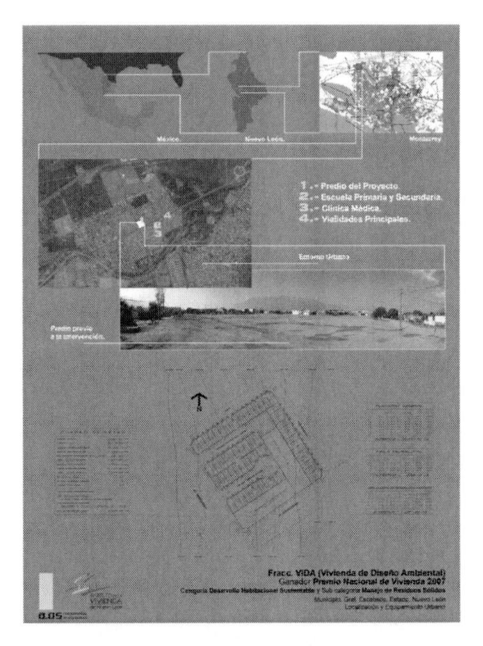

FIGURE 4.1 The location of ViDA in relation to the bioclimatic context. Drawing © DEAR Architects, used with permission.

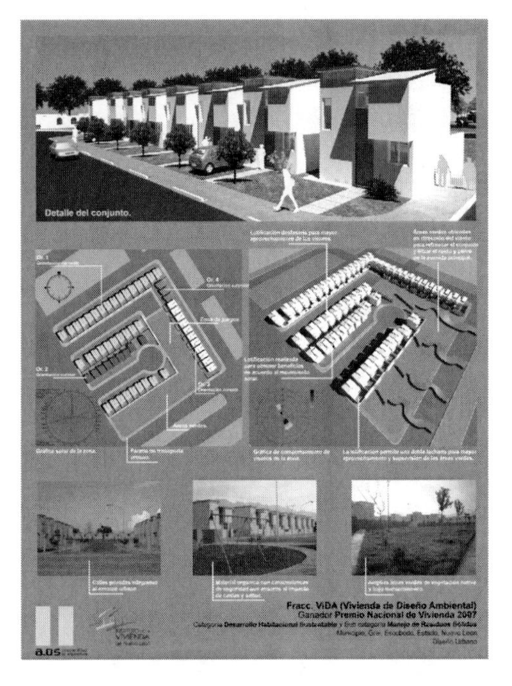

FIGURE 4.2 The layout of ViDA, showing its bioclimatic design. Drawing © DEAR Architects, used with permission.

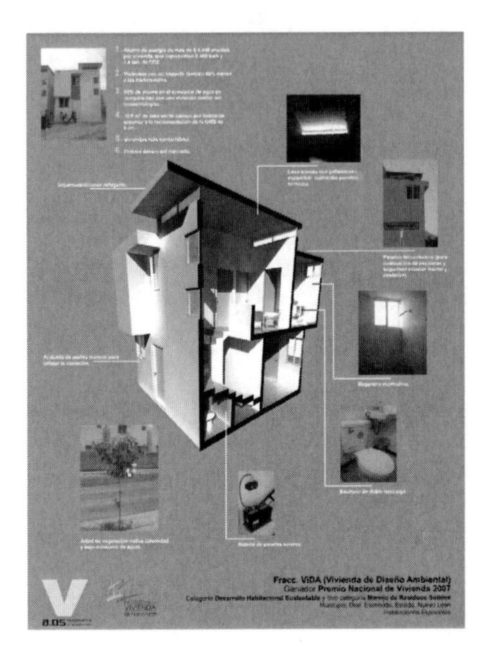

FIGURE 4.3 Features of a ViDA house, showing some of the eco-technologies installed. Drawing © DEAR Architects, used with permission.

bioclimatic components included the study of temperatures, shading and orientation for each house, in order to design the position of each dwelling as well as the insulation, ventilation and lighting systems. Its *eco-technology* components included energy-saving bulbs, water-saving technologies for taps and toilets and solar panels for exterior lighting (Figure 4.3). The architects responsible for the design claim that the measures enable savings of more than US$300 per house per year, also reducing the thermal impact of the houses by 60 per cent and the consumption of water by 50 per cent (with associated GHG emissions reductions), while maintaining market prices for low income social housing. Overall, the project is seen as successful in reducing both the environmental impacts and the costs of utility bills for residents.

ViDA brings together climate change mitigation concerns and concerns about providing social housing for low-income workers, serving to interrogate the notion of addressing climate change and social justice within the context of urban development in the second biggest city in Mexico. As we explore below, multiple actors were implicated in the making of ViDA as an experiment, both in ensuring that it was feasible and compelling. This involved complex processes of calculation which relied not just on the development of new rationalities of climate-friendly housing, but crucially on forging alignment between governmental or semi-governmental institutions with diverging interests: one set of actors focusing on delivering an increasing number of homes and another set of actors focused on improving their quality. In this case, the future residents were not enrolled in the constitution of the experiment, rather it was delivered to them as a complete package. As a result, the project was locally assembled but was neither maintained nor lived as an experiment, such that it has become more or less completely invisible to both the resident and passer-by, as its experimental character has disappeared within the landscape of low-income housing in Monterrey. Yet ViDA's success in supporting a structural top-down transition for the incorporation of eco-technologies in low-income housing in Mexico suggests that dislocations can occur between the success of experiments as sites for innovation within a local context and their power to foster systemic forms of change.

Aligning carbon with low cost housing

A diverse set of actors intervened in the assemblage that made ViDA possible. Forging alignment between competing interests was more crucial than accessing resources or having capacity. But alignment also required the problematization of climate change within housing, such as the objective of making social housing low carbon. Interviewees attributed the success in making the experiment to the confluence of different interests that arose as the result of the coincidence of multiple factors, including the growing interest of regulatory bodies in climate change and sustainability, the availability of financing for housing, the intervention of international organizations and the circulation of new innovations for sustainable housing. The rapid growth of housing within Mexico led to it being problematized as a sector that needed to respond to the demands of sustainability and climate change. Sustainability was to be incorporated within affordability concerns. ViDA's design, for example, was developed in response to a specific instance through which the need for more sustainable housing was articulated: the 2009 call by CONAVI and the Canada Mortgage Housing Corporation (CMHC) (Canada-Mexico Partnership 2010) for sustainable housing pilot projects which emerged from bilateral diplomatic relations between the governments of Mexico and Canada. The Institute of Housing in Nuevo León (an autonomous public organization endowed with its own resources and whose objective is to develop housing policies) was one of the participants. Founded in 2004, the Institute was looking to establish its position as a centre for housing innovation within the State by, among other initiatives, developing a series of pilot projects. Newly recruited personnel at the Institute were also looking for an opportunity to develop innovative bioclimatic models of housing and put in practice skills they had acquired at universities overseas, but which, they felt, were not yet recognized in the existing housing market in Mexico (Interview 5, Consultant, Monterrey, May 2010). The alignment of multiple institutions and individual actors concerned with problematizing housing as a sustainability issue, available resources, and institutional and professional capacities invested in developing forms of pilot response through which to test alternative solutions without disrupting the status quo, served as a potent environment within which to realize the ViDA experiment.

Aside from its avowed concern with sustainable design, the Housing Institute developed ViDA as a conventional social-interest housing project, that is, a project that would not challenge the existing model of social interest housing development. In doing so, the emphasis was both on rendering the problem of sustainability as one that could be addressed primarily through technical means and on forging alignment between the experiment and the existing social-interest housing industry. Central to this alignment was the means through which ViDA sought accommodation with Infonavit. The design of ViDA attempted to provide a sustainable model of housing, which could be financed with Infonavit credits. In order to do, so the design needed to meet two conditions: first, the design should be adapted to the existing housing system standards; and, second, the price of each housing unit had to be kept within the price range deemed acceptable for social-interest housing. In order to achieve the former, the design had to consider the existing context of social housing and the way developers worked with Infonavit. The architect who designed the project explained that developers:

> have their own rhythm and changing their pace of construction is very complicated ... if you want to make any modification is very complicated because they have the model studied deeply in terms of maximum savings ... for example in a block of houses, if the measure is not exact, if the measure is not 4 meters but 4.30, this means that I will spend this more bricks. They have it all very measured, you have to be very precise not to change their parameters.
>
> *(Interview 5, Consultant, Monterrey, May 2010)*

With this in mind, the designers worked on developing housing templates with minimum deviations from existing housing models, while taking into account considerations of orientation, shading, ventilation and lighting. This emphasis on translating the design into the existing templates that would meet developers' demands was achieved at the expense of considerations of the broader structural changes that such an experiment could potentially entail. Working with developers also entailed conforming to the second key criteria for Infonavit credits, that of affordability. The Director of the Housing Institute of Nuevo León explained the importance of maintaining prices for both developers and the workers who bought the houses:

> the builder will say: 'I have to put this, why?' And then the workers say: 'why are they forcing the builder to put a water heater that I will not use and they will charge me for it': the worker has to pay for it, they only thing he will get is more credit. What it needs to be explained is that these measures are taken because they save, that this does not impact the worker and does not increase the price.
>
> *(Interview 3, State Government, Monterrey, May 2010)*

Providing monthly savings to low-income residents by reducing their consumption of energy and water provided the main justification for sustainable housing projects like ViDA. However, when construction started, it emerged that such savings would not be sufficient to cover the additional costs involved in the design of the development. The strategy of the Institute of Housing was to provide a pseudo-subsidy to the developer (a local company called Castor Constructions) by making land available at favourable prices for the project and facilitating the administration and licensing process. As a result, the financial calculations that would ensure that the houses would be eligible for Infonavit mortgages were partly dependent on reducing the capital costs of the project, rather than the financial savings arising from the eco-efficiency measures installed. Once they were completed, the houses sold fast. Residents were attracted to the development because they 'liked the houses' (Interview 6, Resident, Monterrey, May 2010) and because 'they are private and this is more peaceful' (Interview 7, Resident, Monterrey, May 2010). Many were not aware that they were buying into a special kind of sustainable project, even though the project received considerable press attention upon its opening, including the visit of the governor of Nuevo León.

In summary, making ViDA required assembling several heterogeneous components including: technical knowledge that could be developed into a context-led housing innovation model; the need for regulatory bodies to meet policy priorities in sustainable housing as promoted by CONAVI (following the push from Canada-Mexico bilateral organizations and the prominence of climate change in national policy discourses); the independence of the Institute of Housing of Nuevo León to drive forward the pilot project; and, crucially, the availability of resources, first to develop the project and later to subsidize the additional costs encountered in the actual implementation of the project. Simultaneously, making the experiment required working within the existing context of urban development, rather than in opposition to it. In this case, the growth of housing and its problematization in relation to climate change, together with the national social housing programme, allowed an experiment in sustainable housing directed to low-income groups, rather than to high-income residents who could buy into it as a lifestyle choice (see Chapter 3). Here, residents' interests were typified into an ideal citizen who would blindly conform to design demands if they were provided at the right price. Thus, the governing strategy focused on addressing how to shift to low carbon housing with minimum disturbance to the existing social housing regime.

The absence of maintenance

While diverse actors operating at multiple scales intervened in the making of the project, maintaining ViDA was largely left in the hands of the residents, in terms of developing the routine tasks of repair and upkeep that the experiment requires in order to ensure its ability to continue to deliver energy, water and financial savings. In doing so, the experimental character of ViDA was neglected in favour of adjusting its features to accommodate the ongoing circulations in which everyday practices of housing are embedded.

The initial dynamics of upkeep and repair were established through the terms of how the project was delivered and what this implied for establishing maintenance responsibilities. An Infonavit representative stressed 'we have not developed capacities for following up the development after delivery' (Interview 3, State Government, Monterrey, May 2010), a capacity that is allocated to the municipal authority. Strict delivery requirements are established for developments so that the municipality can take care of service delivery without changing their patterns of operation. The representative from the Public Services Secretary in Monterrey explained: 'in new developments, developers provide the pavement, sidewalks, lighting and all, but to our specifications … when we "receive" the new development we have to confirm that it meets our specifications so we can service it and maintain it' (Interview 8, Local Authority, Monterrey, May 2010). This makes municipalities averse to innovation, both in terms of their own practices and in terms of allowing infrastructure and public space innovations in new developments. To minimize maintenance, municipal services will largely rely on forms of service provision which are already in place: they would attempt to minimize the challenges to existing road, lighting, water and sanitation networks from their extension by establishing maximum capacities for such extensions and establishing the form of connection. Equally, when developers deliver the infrastructure associated with a particular project, local specifications will tend to homogenize new technologies with those already existing in the area. The specific regulations, however, are not homogeneous across the Monterrey metropolitan area, with its nine different local authorities with spatially differentiated patterns of growth and service provision. Accordingly, the ViDA model limited innovation to the housing template, rather than communal infrastructures or common spaces (aside from building a small recreational area), following the maintenance patterns of conventional developments. The houses, rather than the surrounding infrastructure including roads, pavements, lighting, water supply, sanitation and green spaces, were regarded as the specific sites for innovation – for architects, designers, policy-makers and developers – and residents were seen as solely responsible for maintaining them. Innovation, in this way, becomes something static, confined to the house which does not directly circulate in terms of fostering deeper infrastructure reconfigurations.

While residents are entrusted with the care and maintenance of each individual house, and thus with the care of the development as a whole, the possibilities of the residents to take care of the home depended upon how the home adapts to existing practices and expectations about maintaining a home. As one resident explained:

> the builder provides a guarantee in case anything happens to the house, for a month, but after that you have to sort everything yourself. They call the house 'bioclimatic'? If that is so, I think it is because it has solar cells, the battery, is charged through the sun. Then you see, the energy is stored and at night you turn on the lights, but the climate, in this study [it] said that if you left these windows open it would help, as if the air is more fluid and there is more air flow but because here the air is already hot in temperature. This was useless. I had to seal them.
>
> *(Interview 9, Resident, Monterrey, May 2010)*

When innovations are adapted easily to residents' practices, like in the case of the solar-powered external lighting, residents are happy to undertake the limited work required involved in their upkeep, and some even lamented that not all the lighting in the house was solar-powered (Resident interviews, Monterrey, May 2010). However, when technologies do not adapt so easily to existing ways of inhabiting a house, for example when the increase of air flow is perceived as heating the house, then residents change them to fit their perceptions of what constitutes a good living environment. Through such practices of upkeep, the features of the houses are adjusted to create an impression of normality among those who interact with the development. However, such forms of upkeep are contrary to the maintenance practices required to ensure that the experiment continues to perform as designed.

Those who promoted the experiment were already aware of the difficulties of relying on residents to maintain sustainability through the life of the house. The representative from the Housing Institute, for example, questioned the use of energy-saving bulbs, one of the features included in ViDA:

> Is it really sustainable to put energy-saving bulbs? If a child with a ball hits the bulb and breaks it and the family goes to the shop to buy a new bulb: will they buy the more expensive one or will they get the cheapest? They will buy the cheaper, it is normal. And then sustainability is over.
>
> *(Interview 3, State Government, Monterrey, May 2010)*

This observation, however, does not correspond with the mixed views found among residents. Sometimes, residents may themselves evaluate the costs of the bulbs against the daily costs of energy. For example, one of the residents explained how she has experimented with different types of bulbs:

> Bulbs? They are energy-saving bulbs. When we first got here we put the normal bulbs and saw that they did increase the spending and we put the energy-saving ones and the difference was clear ... with the normal ones it was 370 pesos and now it is more around 80–90 Well, those are good savings: 300 pesos less per month! You should care about energy. The energy-saving bulbs are sold in malls, they are more expensive, the normal bulb is around 10–12 pesos, the saver is 80. But it lasts.
>
> *(Interview 7, Resident, Monterrey, May 2010)*

However, for other residents, the bulbs were a nuisance because they required additional work of maintenance:

> The bulbs here are quite hard to find [because] you have to go all the way to the centre ... (...) you do not find them easily. You have to go to a big electric shop, you have to take the size. Because besides that here they did not last long, if anything a year they lasted.
>
> *(Interview 6, Resident, Monterrey, May 2010)*

So, while on the one hand the assumption by policy-makers and developers that residents will not care about the technologies of their home is not necessarily correct, on the other hand residents may find it difficult to maintain the experimental aspects of their house if they do entail alternative forms of interaction with their environment or raise the need to penetrate unfamiliar territories such as 'the centre' or 'the big electric shop'. Furthermore, because each resident and each family experiences the house and its environment differently, houses are maintained in different ways. Each house becomes a private project in which normality is to be achieved within the house, rather than being part of the collective project as it was designed.

What the project fails to do is to normalize experimental practices, as the architects and policy-makers who promoted ViDA expected. There is no metabolic adjustment because the experiment is conceived in terms of its integration in a given landscape. But for those who conceived the innovation and would like it to be replicated, towards some form of contextually-based systemic change, there is a need to develop strategic practices that re-experiment the project, that maintain its experimental character. Attempts to re-experiment the project have focused on developing new forms of engagement with its residents. Those who promoted the experiment regard maintenance as a problem of ensuring awareness among residents. For example, one of the architects who designed the project has developed a partnership with the Tec de Monterrey to carry out surveys among local residents which will simultaneously attempt to measure the real consumption in the houses – thus documenting maintenance practices – and raise awareness among residents about how to deal with the ongoing process of repair required to maintain the experiment. Yet such initiatives will have little bearing in the long-term without considering the integration of the house features on residents' life practices, the predominance of a model of private maintenance and the experiential and material aspects of living in these homes.

Living as un-experimenting

The process of living requires the integration of the experiment into residents' experiences and expectations of their neighbourhood and their home. In terms of enacting sustainability discourses, we found that residents do not necessarily regard sustainability as the main objective of the project; rather, they mostly see their homes in the context of their own lives in which concerns about climate change play only a small part. To some degree, the project designers attempted to address these issues by designing low carbon living into the material fabric of the houses, such that residents would not need to be conscious of their engagement with the bioclimatic design. Carbon subjects are here created by assuming that, if the material design of the house is conducive to sustainable practices, residents will conform to them without needing to consciously act upon them, without even being concerned with climate change.

However, different forms of materiality represent points of engagement and tension for residents with sustainable innovations. As mentioned above, one of the key aspects of the development was the natural lighting and ventilation systems, but residents have found this problematic and have chosen to seal the ventilation channels. As one resident explained: 'my husband sealed the windows above, because there was a lot of dust coming inside … the openings are supposedly for light but the air moves them and the dust comes in so that is why we sealed them' (Interview 10, Resident, Monterrey, May 2010). While the architects focused on adapting the houses to the local climate, residents feel the contrary. Furthermore, the combined effect of multiple interventions may itself create unanticipated conditions which escape the initial calculation and which residents themselves are most able to perceive, for example the ventilation system may optimize the temperature inside the house but residents may find that the hot air flow makes lower temperature less bearable in any case.

In addition to the challenges of designing low carbon living into the fabric of individual houses, we found that the ways in which the experiment came to be lived was shaped not only through the use of the houses as homes but also through the ways in which they served as the locus of economic activities and security enclosure. Yet ViDA was only designed to serve as a home space, in which the different home uses were prefigured. Residents were imagined to conform to specific patterns of sustainable living. This represents a deliberate attempt to create subjects to enact low carbon living. Not only were residents imagined as enacting a particular set of life practices, following a nuclear family model with certain temporal dynamics of going to work, bringing children to school and so on, but their spaces of operation were also mapped and simplified. Perhaps the clearest example of a socio-spatial assumption in the design is the separation between the place of residence and the place of work, corresponding to the stipulation that only people in formal employment can access an Infonavit mortgage that would be needed to buy the house. Yet such assumptions failed to recognize the significant variations in terms of livelihoods and household dynamics. Houses are more than simply places of residence; they are shaped by a multiplicity of life activities which can hardly be synthesized in a house template. The effects of these activities are clearly visible in the landscape of the project as it came to be lived: residents shaped their houses as a private project, whether this was with incremental extensions to gain space, adapting the house for a new vehicle or just painting it. In the context of suburban development in Monterrey, two general concerns in particular, each closely related to the circulations which today determine the trends in urban development in the city, have clearly influenced the way ViDA is lived: the need for livelihood diversification; and the context of violence and concern about the drug economy. While residents may only contest the subjectification attempts timidly, their practices end up posing a radical challenge to subjectification.

The need for livelihood diversification means that residents may rely on multiple sources of income. Most residents interviewed praised the location of the development, because it was close to sources of employment in Escobedo, even though it was far from the centre. Yet the development is in a relatively unserviced area. In response, some residents have developed spaces within their houses to provide those services, from child minding to selling tortillas. In some cases, the house has been transformed into a shop. As the architect who designed the project remarks, the result is that:

> there are some houses that are completely modified. They brought down the house and made another one. Why? Because they have a business, because they are so close to important avenues that they brought everything down and built a warehouse where they sell electric material.
>
> *(Interview 5, Consultant, Monterrey, May 2010)*

From building a little cubicle in front of the house, to a refurbishment or even a reconstruction of the house itself, residents' strategies to adapt the houses to existing livelihoods are as diverse as the different ways of making a living found among them. However, such modifications are not as generalized as changes motivated by the desire to securitize the house and the surrounding spaces in response to concerns about the violence associated with the drug cartels in the city. Indeed, residents moved to ViDA in the first place because it was perceived to be safe: 'here is more peaceful ... nobody bothers you here ... but from this corner and further out there is another world ... but nobody bothers you here, you can be with your family without problems' (Interview 11, Resident, Monterrey, May 2010). Yet experiences of violence are common. Another resident explains that there are:

> ... a lot of gangs ... a lot ... you can tell in other developments ... but not inside the development, inside here is not that much, but the moment I go to the border of the development I can see it a lot ... they fight, the houses are vandalized ... that is why we built a bit more towards the entrance because it was quite dangerous: we built as a defence.
>
> *(Interview 7, Resident, Monterrey, May 2010)*

This resident, like three others interviewed informally, had built a high wall around the house 'as a defence', responding to a growing sense of insecurity which is not only related to violence associated with drug trafficking which makes the headlines in the newspapers, but also to the proliferation of violence more generally compounded with the national drug wars. The development of security enclosures, however, also challenges the original template as they clash with the original considerations of shading, ventilation and orientation in house designs (Figure 4.4). This also builds on the tradition of self-building. As one resident explained (Interview 9, Resident, Monterrey, May 2010), many of them have already participated of the culture of self-building, and thus integrate the social housing model with their own attempts at extension, personalization and securitization. The two models of housing hybridize through the living practices in social housing developments.

Furthermore, the making and remaking of houses through self-building practices is leading to certain convergence between ViDA and the surrounding areas, so that ViDA is slowly integrated in the broader urban landscape. What we see here is a

FIGURE 4.4 An example of how ViDA residents have modified their houses. Photo © V. Castán Broto.

breakage from the mould towards enabling new forms of circulation in terms of bringing the experiment in communication with its surrounding areas. In doing so, the boundary of the experiment is undone through the reworking of the assemblage; accretion takes place and a new form of intervention, one which actually enables forms of economic and social circulation, takes root. Here, living practices connect with maintenance needs. While making ViDA required the intervention of multiple actors at different scales and maintaining it was left in the hands of residents alone, through the process of living, residents experiences of the experiment are again inserted in the broader circulations which shape the city's urban development.

Impacts and implications

The paradox of ViDA as an experiment is that, while the project appears to have lost much of its experimental character in the particular urban milieu within which it was developed, it has contributed to substantial change in climate change and housing policies across Mexico. We suggest that ViDA may have intervened in three different ways: first, it has advanced a particular model of incorporating climate change into urban development planning and policies, taking advantage of the existing governance configurations; second, it has provided material evidence for scaling-up a broader sustainability and social housing programme, in the form of the Green Mortgage; and finally, it has established the basis for placing responsibility for addressing climate change and resource consumption on the urban poor, arguably neglecting the roles of those who contribute more per capita to Mexico's GHG emissions, as well as the political and economic interests that serve to structure existing forms of high-carbon consumption.

ViDA has advanced the model of social interest housing guided by Infonavit's stipulations concerning the design and cost of new-build housing. This has had the effect of, on the one hand, demonstrating the feasibility of addressing climate change through housing, but at the same time it has also curtailed the possibility of challenging the inherently unsustainable models of housing which dominate urban development in Mexico. The forms of eco-efficiency that could be accommodated within this model were ultimately not sufficient to deliver housing at a cost acceptable for Infonavit financing, and so the developers had to seek alternative ways of reducing the capital invested in low carbon housing. Furthermore, the ways ViDA residents modified their homes may also have reduced their potential efficiencies, raising questions about the extent to which the aim of accommodating sustainable design within the social-interest housing model was successful at the scale of individual dwellings. The critique that the social interest housing model has led to a particularly aggressive form of urban sprawl, alienating residents from their livelihood sources does not hold in the case of Monterrey due to the polycentric character of the metropolitan area. However, ViDA does little to move away from existing models of urban development in which settlements are developed before services. As the Secretary of Environment and Natural Resources (SEMARNAT) in Nuevo León lamented:

> Sadly, many of these social housing settlements are led by development needs, without planning, in areas without water, without roads, the developer comes, with resources and finance from the government and then when people come to the houses and demand light, transport, schools, shops, they find nothing! We are building everything from the roof down. First we create the settlements, instead of creating services.
>
> *(Interview 12, State Government, Monterrey, May 2010)*

If ViDA largely operated within existing modes of housing development, at the same time it played a small but very significant role in the development of a national-scale Green Mortgage scheme run by Infonavit, which has arguably been much more transformative. ViDA was one of the first developments in Mexico to receive the Green Mortgage, at a time when the requirements of the scheme were still relatively flexible. To be certified for the Green Mortgage, ViDA only had to incorporate water saving technologies, energy-saving bulbs and a solar panel for external lighting. But ViDA's certification was adopted as an example of how the Green Mortgage could be developed and, specifically, how it could be financed and certified. Its significance was reinforced by national-level prizes and other forms of recognition, such as the 2007 National Housing Prize. As an academic who played an instrumental role in developing guidelines for the Green Mortgage explained:

> After my involvement in these projects [ViDA and two others] I started raising awareness of the Green Mortgage and the home certification system and a system of financing. [These ideas] I had already given: 'I have numbers here about how much it costs, in how much time it can be paid, with which benefits and which energetic, environmental and economic systems could be used' because I had it all quantified after that experience …
>
> *(Interview 13, Universities, Mexico City, May 2010)*

Infonavit's Green Mortgage initiative allows loan applicants to borrow more money, where the increased borrowings are used for the installation of 'eco-technologies', such as: water-saving appliances (dual flush toilet, water-saving shower heads, stopcocks); energy-saving appliances (energy-saving bulbs, thermal insulation, eco-refrigerators); and in some cases special measures such as solar water heaters, gas heaters, filters and others. The rationale is that the increased costs of servicing a larger loan are offset by the monthly savings that eco-technologies enable. The initiative has been running since 2007, and over 600,000 Green Mortgages had been written by 2011 when it was made compulsory for Infonavit mortgages. It has been embraced by the construction industry (who benefit from the increased capital available for works), has provided a platform for direct intervention in housing by government institutions that act as certification bodies (particularly the National Commission for the Efficient Use of Energy and the Organization for Energy Savings), and has been used to support the argument that Mexico is a leader on climate change policies. International recognition of the Green Mortgage has come from several international and bilateral institutions, including winning the 2012 World Habitat Award (Building and Social Housing Foundation 2012).

The Green Mortgage has undoubtedly been a success, but ViDA's promoters expressed disquiet about it, lamenting the fact that the Green Mortgage did not follow the principles of bioclimatic design which were so fundamental to ViDA. This highlights one way in which the scaling-up of measures to promote climate-efficient housing has paradoxically led to the ambitions of the experiment being downscaled to fit easy-to-use standardized guides. In a sense, the Green Mortgage is the antithesis of the experiment. The Green Mortgage is not experimental because it moves away from unexplored territories and develops instead as an assemblage of well-known and completely proven standards. Not only does it avoid experimenting with the technologies applied, but it also avoids any form of social, economic or political experimentation, by reproducing existing governance and financial arrangements for social-interest housing. ViDA and the Green Mortgage are in this sense fundamentally different initiatives. ViDA implied risk but transformative potential. The Green Mortgage, by contrast, has a guaranteed impact but its potential is incremental.

Finally, ViDA has had the effect of shifting responsibility for addressing climate change onto the individual and household in a way which has placed increased demands on the urban poor, rather than high-emitters or the political and economic interests who serve to structure existing forms of high-carbon consumption. ViDA produced carbon subjects at the scale of the household, rather than the community, the neighbourhood, the city or the country. Furthermore, the emphasis on resident-led maintenance suggested that individual responsibilities do not end when the worker accepts an increase on the price of the house, but rather continue over time in the collection of practices around the home of repairing and maintaining the different innovations and technologies.

The poor certainly derive benefits from projects like ViDA which seek to financialize carbon and sustainability. However, such projects also draw these residents into new circuits of capital circulation, compelling them to invest, live and act in certain ways. Their ability to access social-interest housing becomes conditional on fulfilling certain sustainability demands on the assumption that these are universal, or that they have a social function. ViDA has also played an important performative role in pioneering the notion of conditional lending, by demonstrating that residents may accept those conditions if they find alternative arguments to access housing, despite the fact that the modifications residents made to their homes diluted the impact of this conditionality in the ViDA development itself. But where the resident is made responsible through a process of subjectification, they inevitably retain some agency, which became clear as the ViDA was adapted to residents' priorities through the process of upkeep, repair and modification they carried out on their dwellings. In the case of ViDA, the effect is that the experimental characteristics of the development are gradually vanishing as the homes are integrated into existing urban circulations, as the modifications to the dwellings gradually bring them into convergence with surrounding areas. Even external attempts to re-experiment ViDA struggle with the contrast between the need to adapt the experiment to residents' expectations and the belief that residents can follow idealized models of experimental practice. This stands as an example of the fact that assembling an experiment is by itself insufficient for a broader reconfiguration of the urban landscape.

References

Building and Social Housing Foundation (2012) *World Habitat Awards Announced!* Available online at http://www.bshf.org/ (accessed, 26 August 2012).

Canada–Mexico Partnership (2010) *Annual Report.* Co-published by the Government of Mexico and the Government of Canada. Available online at http://www.economia.gob.mx/files/Canada_Reporte_CMP_2009.pdf (acessed November 2013).

Federal Government of Mexico (2013) *National Climate Change Strategy.* Available online at http://www.encc.gob.mx/en/index.html (accessed November 2013).

Infonavit (2012) *Infonavit en cifras*. Available online at http://portal.infonavit.org.mx/wps/portal/EL INSTITUTO/Infonavit en cifras/ (accessed 22 August 2012).

Montes, X. and Ortega, X. (2003) *Marginality Indicators in the Monterey Metropolitan Area*. CDEM.

Nuevo León Governmen, (2012) *Population in Nuevo León*. Available online at http://www.nl.gob.mx/ (accessed 28 August 2012).

5

THE 'COOLEST BLOCK' CONTEST IN PHILADELPHIA, USA

Introduction

In this chapter, we examine the *Retrofit Philly Coolest Block Contest* (Coolest Block contest) – a competition run in 2010 which gave residents of Philadelphia the opportunity to win a cool roof, air sealing and insulation upgrades by joining together and submitting an entry for their block. The contest was sponsored by the Energy Coordinating Agency of Philadelphia (ECA), Dow Chemical Company's Building and Construction Division (Dow) and the City of Philadelphia Mayor's Office of Sustainability (MOS). We explore the Coolest Block contest and its implications for addressing urban sustainability, infrastructure renewal and climate change in the City of Philadelphia. A port city on the Delaware River at the centre of a metropolitan region of approximately 6.3 million people, Philadelphia has always been a centre of commerce. From the late eighteenth century it was one of the most important ports on the East Coast of the American colonies, and it developed into a key industrial powerhouse in the early twentieth century with the development of rail links into the city. Rapid urban development followed industrialization in the form of rows of terraces constructed to provide housing for the burgeoning industrial workforce. However, Philadelphia's fortunes turned with de-industrialization and decentralization after the 1950s. Factories closed, population declined and land lay vacant and, as middle income white households in particular moved to surrounding counties (City of Philadelphia 2011), poverty became increasingly racially segregated, with black residents concentrated in the most blighted city centre neighbourhoods. In this context, urban sustainability and particularly climate change have come to be seen as a new means through which the city might promote infrastructure renewal, redevelopment and economic growth.

The Coolest Block contest epitomizes the ways in which the climate change agenda, broadly conceived, is being put to work in relation to the broader economic and social challenges facing cities. In this chapter, we examine how the Coolest Block contest became embedded in the wider circulations of the city, enabling action on urban development issues which were in many ways quite unconnected to either climate change or sustainability agendas, such as improving the housing stock in the context of high poverty rates. The case suggests that authority can be extended to realms other than those within which the experiment is assembled. Examining the wider context of experimentation in Philadelphia and how the Coolest Block contest was made, maintained and lived, we show that the will to improve that is pursued through experimentation creates heterogeneous geographies of benefit. In Philadelphia, this allows those with existing capacity to lay claim to the benefits of lower carbon homes, whilst also facilitating corporate benefits in the form of an emergent market for the expertise and technological solutions of companies such as Dow.

Climate change and cities in the USA: agenda setting in Philadelphia

The USA has been far from a leader on climate change policy at the national scale, notably never ratifying the Kyoto Protocol and playing a largely obstructionist role in the international climate change negotiations. However, at the same time, states, cities and regions within the USA have signed up to a range of actions to mitigate the effects of climate change. For instance, Byrne *et al.* (2007) observe that 19 per cent of the US population live in cities which have signed up to the ICLEI Cities for Climate Protection campaign, and some 435 US cities pledged in the 2005 US Mayors' Climate Protection Agreement to meet or exceed the proposed emissions reduction target for the USA under the Kyoto Protocol. A small number of influential cities have been notable for their attempts to link the agenda of climate change to urban sustainability agendas, most notably New York, Chicago, Seattle and Portland. In the context of economic austerity and the need for urban revitalization, Philadelphia has sought to position itself in as a leader in urban sustainability as part of this movement. In 2007, the City committed to the Cities for Climate Protection Initiative, the US Mayors' Climate Protection Agreement and the Large Cities Climate Leadership Group and Clinton Climate Initiative. Despite this action, the Pennsylvania Environmental Council (2008: 4) argued that 'the greening of Philadelphia is lagging far behind that of other major American cities like New York, Chicago, Washington, and Seattle'. In the run-up to the 2007 Mayoral election, mayoral candidate Mayor Nutter differentiated himself from the other Democrat contenders by embracing sustainability, and when he was elected one of his first actions was to establish a Mayor's Office of Sustainability (MOS) in January 2008, initially with 2–3 full-time employees. The first task of the MOS was to draft a new sustainability plan for the city, which it released in 2009 as *Greenworks Philadelphia* (City of Philadelphia 2009).

Greenworks was created to 'help the city leverage its existing assets and mitigate its exposure to the effects of global warming' and focused on both 'changing the way that government does business' and 'giving citizens the tools they need to lower their own carbon emissions and reduce their vulnerability to increasing energy costs' (City of Philadelphia 2010a). Organized into five themes (energy, environment, equity, economy and engagement), the plan set aspirational targets and milestones, such as the first goal, which was to reduce energy consumption of the City by 30 per cent by 2015. Though innovative, *Greenworks* drew on ideas that had been developed in Chicago and New York (Interview 1, Municipal Authority, Philadelphia, October 2011) and has been heavily reliant on both partnerships and the massive injection of federal government funding provided by the *American Recovery and Reinvestment Act 2009* (ARRA) in order to deliver its objectives. ARRA was an unprecedented federal government intervention in the economy, designed to stave off the most severe implications of the 2008 global economic crisis. The overall value of the package was US$787 billion (subsequently increased to US$840 billion), of which US$3.1 billion nationally was allocated to be spent on energy efficiency measures. Pennsylvania received US$99.6 million funding to spend in a three and a half year window, in addition to US$33 million from the Department of Energy's Energy Efficiency and Conservation Block Grant (EECBG) programme, creating investments of orders of magnitude greater than that available from regular funding (Interview 2, State Government, Harrisburg, October 2011). For Philadelphia, the significance of ARRA funding cannot be understated, as a respondent from the ECA reflected:

> the city really took full advantage of the ARRA funding – the stimulus money – and they were definitely in the right place at the right time with planning … Philly not only had a lot of you know kind of big plans in place but they actually had you know kind of started figuring out the implementation of big chunks of this work and could deliver this work in the timeframes, these tight timeframes that the stimulus money had.
>
> *(Interview 3, NGO, Philadelphia, October 2011)*

One such programme was *EnergyWorks*, launched in 2009 using US$25 million in ARRA funding and designed to achieve the second goal of *Greenworks*; namely to reduce city-wide building energy consumption by 10 per cent. *EnergyWorks* is a one-stop-shop which provides a low-interest revolving loan programme which provides loans of US$5,000 to US$15,000 to finance residential energy efficiency works (Interview 1, Municipal Authority, Philadelphia, October 2011) and connects applicants with contractors to carry out the works. It also provides loans of US$100,000 to US$1 million for commercial projects.

The confluence of growing momentum at the municipal level in the United States for a response to climate change, the election of a Mayor who wished to demonstrate leadership in this area to differentiate himself, and the injection of substantial levels of capital investment as a response to the economic crisis of 2008 created a potent context for action. In this context, long-standing issues of energy efficiency and the retrofit of the residential sector of the city came to be seen as both a problem in need of intervention and an arena in which economic and environmental benefits could be secured. This gathering momentum led to the formation of new alliances and programmes designed to reach into the domestic sphere, predominantly through *EnergyWorks*, and to the formation of the specific initiative which is our concern in this chapter – the Coolest Block contest.

Partnership, competition and community in the Coolest Block contest

The Coolest Block contest was a competition run in 2010 which was designed to help meet Target 3 of *Greenworks*, namely to 'Retrofit 15 percent of housing stock with insulation, air sealing and cool roofs' (City of Philadelphia 2010b: 5). It aimed to demonstrate the potential of energy-efficiency retrofits including cool roofs, air sealing and insulation as a means of both saving energy and improving the quality of the city's housing stock. Philadelphia has a distinctive housing stock made up predominantly of flat-roofed terraces which were first constructed to provide cheap housing for the growing industrial workforce in the eighteenth and nineteenth centuries. Often constructed from poor quality materials, these 'Philadelphia Rowhouses' are subjected to the extremes of both heat and cold experienced in Philadelphia, leading to high energy bills for heating and cooling as residents seek to achieve thermal comfort. The contest aimed to mobilize the community by block, consisting of the homes occupying both sides of a street between two cross-streets, and 76 blocks entered the competition, which involved a petition in which homeowners agreed to participate, and a 500-word essay on why their block was the coolest block in Philadelphia. Upgrades were installed on the winning block – 1200 Wolf Street – between July 2010 and May 2011 (Figure 5.1).

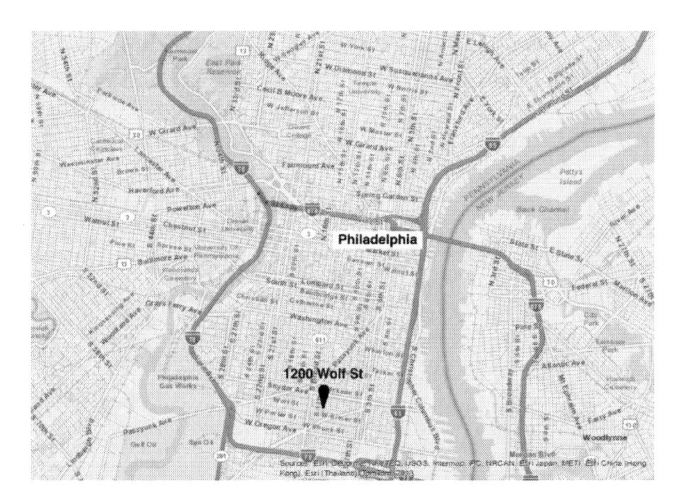

FIGURE 5.1 Map of Philadelphia showing the location of 1200 Wolf Street. Drawn by V. Castán Broto on ArcGIS Basemap.

In contrast to more mainstream approaches in which government agencies seek to encourage energy efficiency retrofitting through offering financial incentives to address the capital costs or payback periods for the installation of different technologies or to enact behavioural changes amongst households, the Coolest Block contest was innovative in bringing not-for-profit, municipal and private sector partners together, seeking both a collective response amongst neighbours and one that positioned different blocks as competing with one another for the resource available. In stressing the potential of alternative roofing materials for cooling residential buildings, the contest also sought to test the viability of such materials and to make them visible within the city, thus developing forms of calculation that would enable climate change governance. As we outline below, the initial opportunity for the contest emerged from the new alignment of interests brought about by the establishment of the *Greenworks* programme and the political investment that accompanied it, but even in this context considerable strategic work was required to assemble the experiment.

From serendipity to strategy: making the Coolest Block contest

In 2008, Mayor Michael Nutter convened a task-force of leaders from government, NGOs and the private sector to sit on a new Sustainability Advisory Board in support of his vision to 'make this city the greenest in America' (City of Philadelphia 2009: iii). The 21 members of the Sustainability Advisory Board were to provide advice and support for a major new sustainability initiative being drafted by the Mayor's Office of Sustainability, which would ultimately be launched in 2009 as *Greenworks Philadelphia*. Sitting on this Board were a representative from the ECA and a representative from Dow who had experience working together, and their personal connection opened up the opportunity to take advantage of a specific and temporally limited alignment of interests between the three organizations which became the project partners (Table 5.1). As Interviewee 3 put it:

> [Interviewee 4, from Dow] came to me for ideas and I said well why don't we do a Coolest Block contest because we had done one of these before and it was a huge success ... and so [Interviewee 4] loved the idea.
>
> *(Interview 3, NGO, Philadelphia, October 2011)*

The ECA's desire to promote white roofs emerged from its experiences with weatherizing low-income homes, where leaking roofs regularly prevented meaningful energy efficiency retrofits from being installed. White roofs, in the minds of the ECA, were a retrofit which overcame this barrier by providing both the social benefit of a structurally sound, water-tight roof and the environmental benefit of reduced energy use for cooling and therefore reduced greenhouse gas emissions (Interview 3, NGO, Philadelphia, October 2011).

At the outset, the Coolest Block contest was more a product of serendipity than strategy. While certain strategic actions were performed, forging alignment

TABLE 5.1 Overview of the partners' objectives for and contributions to the Coolest Block contest

Partner	*Objectives*	*Contribution*
Energy Coordinating Agency (ECA)	• Promote energy efficiency in Philadelphia	• Ran the contest • Project managed retrofits
Dow Chemical Company	• High visibility signature project to commemorate and celebrate the unification of Rohm & Haas and Dow in 2009 • Demonstrate the energy savings its products could deliver, and the fact that it as a company took sustainability seriously	• Project funding of *c.* US$500,000 from the Dow Foundation • Supplied the products for the retrofits • Technical support and materials expertise • Publicity materials and media engagement
Mayor's Office of Sustainability (MOS)	• Make visible progress on *Greenworks*, particularly the third target which was to 'Retrofit 15 percent of housing stock with insulation, air sealing and cool roofs'	• Administrative support and publicity

was only possible because of a corporate takeover, the election of a new Mayor (one of whose platforms was action on sustainability), and the pre-existence of an active and well-connected not-for-profit organization that was ready to mobilize at that precise moment. It also depended on a global financial crisis that inspired the federal ARRA funding package. A gathering momentum to address climate change and energy issues beyond the context of Philadelphia which drove each of these actors served then to align the interests involved in the city, who then applied considerable strategy to implement the project. For instance, participants for the contest were recruited by a MOS mailout of a glossy information pack (supplied by Dow) to the Mayor's network of Block Captains, which Interviewee 3 described as:

> people who are either self-appointed or you know somehow elected or selected by their neighbours as the leader of the block … so the block captain is the point person for those civic engagement kind of issues and the mayor has cultivated that group of people very wisely and he maintains a very good relationship with block captains.
>
> *(Interview 3, NGO, Philadelphia, October 2011)*

In this sense, the contest was very dependent on a form of engagement with the community which was both specific to and proven as effective in Philadelphia. This engagement strategy was also taken up subsequently in *Greenworks*, which involved enrolling citizens in action through local networks providing (at least the perception of) direct access to powerful figures such as the Mayor.

This initial engagement was backed up with three community information evenings to provide further information on the contest and the materials which would be installed on the winning block. This gave potential entrants an opportunity to:

> touch and feel the materials, they could see them ... and the people from Dow were there so they could see that you know the money was real, this wasn't fake you know and Dow was really going to put this money out there.
> *(Interview 3, NGO, Philadelphia, October 2011)*

This rendered the contest technical by emphasizing the materiality of the retrofits being offered. These encounters also ignited the competitive nature of the contest:

> they saw each other and they were like oh this is a real contest you know that group over there you know I don't want them to win you know I gotta win you know so it started this whole thing going about the competition and winning.
> *(Interview 3, NGO, Philadelphia, October 2011)*

In this way potential participants were enrolled in the contest – they had to demonstrate interest, but in return were shown the potential benefits, while at the same time the competitive nature of the initiative was emphasized. The making of the experiment thus required both assembling common interest amongst the organizers and a process by which the contest was made compelling to the broader population.

In addition to installing a cool roof, air sealing and insulation upgrades, the winning block would have a block party thrown for it, which would be attended by Mayor Nutter. Owners agreed that, if successful, they would allow PECO and PGW (the electricity and gas utilities) to supply one year of electricity and gas consumption data respectively to the contest organizers, and Dow would also undertake long-term monitoring of ambient air temperatures at the block. Participation in the contest was limited to blocks that were primarily residential, and 76 blocks subsequently entered the contest, most of which were from either West Philadelphia or South Philadelphia (Interview 3, NGO, Philadelphia, October 2011). Dow's representative thought that there would have been more entries but that the perceived inevitability of who might win had put off a number of participants:

> the block captains do talk to each other and Wolf Street had like just about a hundred percent [participation]. So when people heard that they had it signed off a hundred percent, if you didn't have a hundred percent you know ...
> *(Interview 4, Corporation, Philadelphia, October 2011)*

In the end, only one successful block was required for the competition to be considered a success, so the processes of rendering the experiment compelling were directed towards this end, with the effect that the contest perhaps encouraged apathy at the same time as fostering active participation and engagement for particular communities.

Participants were told in a fact sheet about the contest that the criteria for assessing entries would 'include the condition and type of roof currently installed, the level of participation for your block, and the story of why your block should be chosen as the "Coolest Block in Philly"'. Though neither the scoring system nor the short-list of blocks it generated were made public, interviews with the organizers revealed that the criteria were mainly set by Dow. In doing so, Dow was able to deploy a particular calculation which aligned the contest with its own interest, without openly conflicting with the concerns of its partners. The first and most important judging criterion was high participation, because Rowhouse roofs are connected along the length of a block and the elastomeric white roof coating is applied in a similar fashion to spray paint, such that all households needed to participate to make it efficient and effective. Second, the roof had to be in good condition, as Dow did not want to have to spend large amounts of money on making good existing roofs before the white roof product could be installed (Interview 3, NGO, Philadelphia, October 2011). Third, it was advantageous if there were a high number of homes on the block. There was no income criterion used in the contest, and the essay was included in the judging criteria after consulting an attorney specializing in contests, who advised that the inherently subjective nature of judging an essay would reduce the likelihood of losing blocks challenging the conduct or result of the competition. This was successful to some extent. One resident from the winning block reflected that 'My husband just said to [the block captain]: be truthful' (Interview 5, Resident, Philadelphia, October 2011) in the essay, which she thought was a reason that 1200 Wolf Street won the contest. In actual fact, the essay was the least important judging criterion, but did have unexpected benefits, providing an important snapshot to the contest organizers of the concerns of Philadelphians, in particular 'that many, many Philadelphia home owners were worried about losing their homes due to the rising cost of energy' (Interview 3, NGO, Philadelphia, October 2011).

Discussions with residents from the 1200 Wolf Street block – which won the competition – suggested that they understood the strategic importance of the criteria. The block captain explained that from the earliest community meetings about the Coolest Block contest, she sensed that getting a high a level of participation amongst residents would be the most important criterion in deciding on the winner, and so she spent a lot of time door-knocking her block to get support. She also thought that the condition of roofs would be important, since Dow had shown its concerns about them (Interview 6, Resident, Philadelphia, October 2011). From her perspective this placed 1200 Wolf Street well, because she knew it was a well-maintained block. Another resident (Interview 7, Resident, Philadelphia, October 2011) corroborated this insight. But the block captain also sensed that the organizers were keen to find a winner from South Philadelphia 'because it's the area of the city with the least green space, it's the biggest heat sink in the whole city' (Interview 6, Resident, Philadelphia, October 2011). In this sense, the contest explicitly problematized the issue of urban green spaces through a climate change lens, and sought to act on a so-called typical Philadelphia block as a demonstration of the possibility of the retrofits installed to have a wider impact in the city. The 1200 block of Wolf Street is in many ways stereotypically South Philly, with its high proportion of Italian American residents and tight-knit community. But the coloured lights strung across the road from rooftop to rooftop serve as a visual marker of the fact that it is also exceptional (Figure 5.2). It is notable for having a low proportion of renters, and one long-term resident commented on the strength of community on the block: 'this block you wouldn't be able to get into, it was so strong' (Interview 5, Resident, Philadelphia, October 2011), making it a desirable address in South Philadelphia. The block captain, who had bought her house much more recently, also noted that it was a neighbourhood on the rise:

> the neighbourhoods just north of that are really popular areas to live the last few years and it's like it keeps creeping further south. So on our block there's a good mix of people that have been there their whole lives and people who are thirty-five with two kids and been there for a few years.
>
> *(Interview 6, Resident, Philadelphia, October 2011)*

The block captain of 1200 Wolf Street was also an important actor in her own right. She described herself as an environmental engineer by profession and former project manager, with an architect husband. This proved important both to the success of the block in the contest and to the success of the project as it moved

FIGURE 5.2 The 1200 Wolf Street block; the 'Coolest Block'. Photo © G.A.S. Edwards.

into the construction phase. The block captain was clear that she was keen to participate because of the personal benefits it would bring:

> I was definitely interested in giving it a whirl both from you know what I would gain personally as a home energy from the energy efficiency upgrades and also again conceptually from seeing would that really work to like put a white roof on an entire city block what kind of an effect would that have.
>
> *(Interview 6, Resident, Philadelphia, October 2011)*

This gave her the motivation to undertake the hard work of selling the contest to the other residents. She reflected 'me and my two oldest [children] at the time walked for like three weeks – it was like every day … lots of cups of coffee', but this effort was rewarded by the fact that:

> there was relatively little resistance … I think that is very indicative of our block, I got a lot of, you know 'oh well if it's going to be something good for the block then I'll sign it', they almost didn't care what it was going to entail.

The importance of the role of the block captain in enrolling her neighbours into the project was not overlooked by either the contest organizers or her neighbours. One block resident reflected that 'she basically did it and she won it, and they – the Dow people – came around and they did the work' (Interview 5, Resident, Philadelphia, October 2011). The organizers, likewise, explained that the block captain 'spoke to everyone and she tailored her pitch to them based on what she knew about them right whether because she's, she's what, an environmental scientist? … And her husband's an architect' (Interview 4, Corporation, Philadelphia, October 2011), and that:

> I think forty-seven out of forty-eight properties agreed to participate which was pretty far above everything else and I think just a testament to how hard the block captain worked to get everybody on board because people are inherently suspicious about the city government in some parts of the city.
>
> *(Interview 1, Municipal Authority, Philadelphia, October 2011)*

They rewarded this effort with additional insulation and air-sealing retrofits to the exterior walls and ceiling of her house's front porch, which also forms the floor of the front upstairs bedroom.

This analysis suggests that, as it moved from emerging in the serendipity of the new alignments being forged around sustainability to the strategy of aligning the project, the making of the Coolest Block contest relied on the contest organizers mobilizing the subjectivity of community in order to create discursive room for the project and gain access to the territories (peoples' homes) on which they desired to act. The will to improve was perhaps not fully deployed at the outset of the project, when alignment was the result of a confluence of circumstances, and yet, as the experiment progressed, it was manifest in the interventions of different actors who came with a series of calculative practices about criteria for selection (Dow) or estimating what those criteria were (Block captain). The efforts required to deploy such strategic tools within a particular socio-technical assemblage required an active engagement, which in no way can be seen as a product of serendipity. The problems of climate change adaptation and mitigation were transformed through their engagement with communities into problems of comfort and potential economic advantage, and whilst this was successful in motivating participation in the contest, it also meant that only the block captains were forced to engage with the underlying rationale for the upgrades, and could selectively choose different narratives to present to their neighbours so as to maximize participation. As a result, after the retrofits had been installed on the winning block, some residents still seemed unaware just who was involved in the competition or what their interests were, an issue to which we return below in considering how the experiment came to be lived. The contest was made compelling to only a small number, and this would affect both the success of its maintenance and its ability to transform the lives and practices of the residents for whom it was established.

Maintaining the experiment within community: the role of intermediation

Maintaining the momentum of the Coolest Block contest after the winning entry was announced required both processes of upkeep and metabolic adjustment (Chapter 2). On the one hand, upkeep was required to ensure that the installation of the retrofit proceeded, and, on the other, metabolic adjustment was required to widely circulate the potential benefits across the city and also to enrol others in the broader *Greenworks* goal of retrofitting. The upkeep of the experiment in this case relied on the particular mix of professional expertise and community embeddedness embodied by the block captain, which enabled her to act as a vital intermediary between the organizers and the residents as the contest was transformed into a construction project. She quelled resident concerns that the City would always attach strings to benefits provided by the contest, and helped them overcome unease about having contractors actually enter their homes to carry out the works. At the same time as facilitating the organizers' access and her ability to undertake the construction works, she saw herself as the advocate of the block's interests to the organizers. For instance, insulation installations required cutting a hatch in the roof, with potential implications for the warranty of existing roofs, and residents needed documentation on the roof coatings and information on what warranty, if any, they would receive. The block captain reflected:

> I don't think the ECA was entirely ready for [my husband] and I either because we were critical of course of project management and ... I understand you can't give us a warranty for something that we didn't pay for but you know what if people go to sell their house in the next ten years what do they tell the new owners about the roof and you know what if they are

> still there in ten years and they want someone to recoat it. They need to know exactly, they need some sort of documentation of what it was coated and with what so that the next roofer knows how to do it correctly … I don't think they expected two people to have construction management backgrounds to have won the contest [laughs] and be very you know I took it very seriously as my responsibility to be you know the advocate of the people on the block as well as the go-between with the ECA.
>
> *(Interview 6, Resident, Philadelphia, October 2011)*

The experiment required active intermediation from within the block to legitimize it and allow it to transform the material dwellings of the block. Critically, the professional experience of the block captain and her husband allowed them to reconcile the aspirations of the contest organizers in terms of the retrofits to be undertaken with the constraints apparent only to residents – mundane but important questions about documentation, warranties and liability for instance, any of which could have prevented the installations going ahead.

At the same time as realizing the project through its implementation, the organizers were particularly concerned to portray the contest as a success story. On 19 June 2010, four months after he and a representative from Dow had launched the contest, Mayor Michael Nutter celebrated with residents of the winning block and representatives from the contest sponsors with a street party. Media were in attendance and a cherry-picker gave them the chance to see the new white roofs of the block for themselves. The publicity generated by the contest was seen as invaluable in keeping up the appearance of the contest and ensuring that it became visible to multiple communities, because most of the retrofits it promoted are invisible both from the street and to residents, located as they are on rooftops which are not necessarily visible from street-level or within the building lattice, inside walls and ceilings. So extra maintenance was required to keep the retrofits in the field of vision of both the public and particularly elites, in the hope this would stimulate processes that would lead to the wider adjustment of flows of resources across the city, generating metabolic adjustment. To this end, at the end of August 2011, an educational field trip for a busload of 20 congressional staff members for environmental initiatives included a tour of 1200 Wolf Street, and the ECA hoped this experience would lead them to adjust a formula measuring the ratio of benefit, so that white roofs would meet the payback period specified by legislation. To maintain the experimental character of the contest, it required ongoing processes of ensuring that it was visible to worthy audiences, through trips, surveys and other methods of re-experimentation. However, the PR effort generated by the contest was not consistent. Though Dow in particular produced a range of PR materials including videos (available on YouTube) brochures, T-shirts and other accessories (Figure 5.3) there is no evidence of further effort to engage with the other blocks which had entered the contest once the winner had been selected, apart from inviting them along to energy education workshops, which can be seen more as an effort of subjectification than of experimental maintenance.

FIGURE 5.3 A promotional T-shirt for the block party run at the end of the Coolest Block contest. Photo © G.A.S. Edwards.

Long term, largely at the behest of Dow, the organizers hope to use the 1200 Wolf Street block to show how the retrofits have saved energy and improved comfort for residents through analysis of energy bills and monitoring of ambient air temperatures, to demonstrate the benefits of such retrofits in achieving both financial and environmental goals, and show the credentials of Dow's products. However, even in the absence of this analysis, the organizers were at pains to construct a narrative in which the contest and the retrofits it provided were shown to be a success. For instance, the fact that one house did not get the cool roof coating was seen as a benefit, since it would only make Dow's monitoring of energy use more interesting. Deploying narratives of success is also a key factor to make maintenance over time possible. The representative from Dow explained that the organizers were trying to write a paper about the contest to publicize it, Dow was working with National Renewable Energy to use the lessons they learned from the Coolest Block contest in other retrofitting projects, and the MOS was planning to use the block-scale retrofit approach adopted in the contest in future projects funded by the energy efficiency block grants from the US Department of Energy.

Not only were the retrofits themselves consistently promoted as a success by all involved, but equally the value of the novel form of partnership working involved was also frequently reiterated. The ECA reflected that:

> Dow … were just absolutely delightful to work with. So this kind of project is difficult to replicate so because it's kind of a … one of a kind project but what we're seeing is the PR, the public communications that was done and has been done around our white roof efforts has really gotten white roofing established in the city.
>
> *(Interview 3, NGO, Philadelphia, October 2011)*

A representative from the MOS, likewise, argued that it 'was I think a really big success and it worked out really well in terms of the timing that we had a cool roof legislative effort happening right around the same time as the competition' though he acknowledged that this was a timely coincidence. The representative from Dow reflected 'from my point of view it's been fabulous. Why? Because one, people are still talking about it. Two, even better, they're doing something about it' (Interview 4, Corporation, Philadelphia, October 2011). She went on:

> And people saw it and they loved it. I mean what's not to love about it? I mean it's really the Philadelphia story, it's communities, it's how they all got together and they didn't know what all this stuff was but I mean you can see the, if you go to [the ECA] website right, they have the how you do a blower door test and how you do the smoke test and you know you have these people who sit out on their porches and all talk to each other.
>
> *(Interview 4, Corporation, Philadelphia, October 2011)*

The Coolest Block contest was temporally bound – happening at a particular moment in time – but the actors who promoted it were keen to maintain this alignment of interests beyond the contest, to give it longevity. They did so by continual reproduction of the benefits of the project and its compelling elements, as the representative from the MOS explains, one in which 'everybody who participated has been thrilled with it and it's nice, it's a great photo to get down there and see a whole block both sides that's covered in white roofs' (Interview 1, Municipal Authority, Philadelphia, October 2011). However, once the picture was taken – the contest run, prizes allocated, and successes celebrated – the actors involved found that they needed to institute new processes to perpetuate the idea of the contest, which would otherwise fade from consciousness.

However, it was clear that the benefits from the contest were unevenly spread even amongst the contest organizers. Dow saw both internal and external benefits. Internally, the contest demonstrated to former Rohm and Haas employees in Philadelphia that 'Dow were here and here to stay' (Interview 4, Corporation, Philadelphia, October 2011), as well as building links between a number of Dow's business units and between people from its Midland Michigan headquarters and Philadelphia.

At the same time, externally, it facilitated an exchange of views with community members on the potential applications of Dow technology. For Dow, the block party:

> was a chance to really just sit around and talk to people about what it [the retrofit process] was like and just to hear about and I think for us scientists it's really good to hear well what did they like about it? What was hard to do? What was easy to do?
>
> *(Interview 4, Corporation, Philadelphia, October 2011)*

Perhaps most importantly, the contest created a new alliance between Dow and the City of Philadelphia. As the representative from Dow explained: 'at the city I mean the mayor can't wait to do another project with us [laughs] and I can't wait to come up with another project to do'. The strategic forms of calculation deployed during its assemblage thus directly influenced how different actors could benefit from the experiment.

A representative from the coatings company Acrymax was more circumspect on the benefits of the contest. Acrymax had been involved as manufacturer of the elastomeric coatings used on the roofs, using resin technologies and materials from Dow. Though he thought the Coolest Block contest was a successful project, he reflected that it was challenging to translate pilot schemes like this into ongoing business, because they are inserted into the urban fabric at a particular place and a particular time, alignments which are hard to maintain. He asked:

> where does it become a point where it's a jobs creator and you're actually employing more people because half the City of Philadelphia is installing cool white roofs with our coatings and you're making money on it? … Where do we get to further our business plan?
>
> *(Interview 8, Corporation, Media PA, October 2011)*

For him, this was not just a business challenge, however. He explained that Acrymax had been involved in installing white roofing on:

> five or six hundred homes in the City of Philadelphia … but on the grand scheme of things these five hundred homes or six hundred homes in the City of Philadelphia it's a drop in the bucket. To really do it on a sustainable level and go to that level you need to get the whole like society.
>
> *(Interview 8, Corporation, Media PA, October 2011)*

Furthermore, while the goal of the contest was to promote neighbourhood retrofit, in his experience the retrofits it had inspired were mainly single dwelling retrofits, which could not take advantage of the benefit of economies of scale for both clients and contractors. Here we encounter an implication of the contest's transitory nature: the maintenance required to mobilize the contest in broader urban transformations proved more challenging than the process of assembling the contest and promoting its successes. The idea of the contest was most notably mobilized in support of the *EnergyWorks* programme, which provided low-interest loans to fund energy efficiency improvements, but though the contest provided a great photo-opportunity for launching the new legislation to mandate the use of cool-roofing on new construction in the City (discussed in more detail below), it seems that it had much less success in stimulating the take-up of cool roofing. In part, this may be because of a lack of ongoing support for public participation in retrofitting in the form, for example, of incentives, and reliance instead on the efforts of industry in opening up a new market for retrofitting on the back of the wave of publicity the contest created.

Living low carbon in Philadelphia: subjectification and agnosticism

While the contest was widely celebrated and was actively put into circulation amongst the professional networks connected with its inception and deployment, the extent to which it became actively taken up in the daily practices of either the residents of the winning block or within wider communities is rather less clear. Within the block, it was clear that, though the residents welcomed the improvements they had received, they did not necessarily connect them with climate change, but rather with comfort or economic gains. The block captain, for instance, explained that her gas and electricity usage was almost the same between the two summers immediately before and after the contest, but noted that the arrival of her third child in February 2011 meant a direct comparison was impossible. When her roof was coated in early July, temperatures had already been reaching 90°F (32.2°C) and she found it was noticeably cooler upstairs at night the first night after it was coated. Another resident (Interview 5, Resident, Philadelphia, October 2011) was very positive about the contest, but said that she did not have to have much done to her house because her husband had done most things after they bought the house around three years ago: 'the house was done; they didn't have to do much to my house'. Despite initial resistance from her husband to the white roof coating, they had agreed to it in the end and she thought it had saved energy. She estimated that her energy bill was 30 per cent lower after the retrofits than the year before because 'insulation saves you money'. By contrast, another resident thought that 'the air-conditioner definitely runs less' since the contest, but was not sure that there was much of a difference in electricity bills. Unlike her neighbour, she was not sure about climate change, saying that there had been hot summers in the past: 'some summers are hot, some aren't' (Interview 7, Resident, Philadelphia, October 2011).

While the block residents were largely passive in the pursuit of the contest and the strategy employed by the block captain to ensure that Wolf Street was successful, this range of responses illustrates the ways in which the contest had come to be regarded as part of the identity of the Block and normalized within the everyday responses of at least some of the residents. In making the energy retrofitting part of the daily fabric of their lives, however, climate change remains for most a distant concern. This spread of opinions reflects the fact that the ECA and Dow downplayed climate change and emphasized personal benefit in promoting the contest, a framing consistent with that observed in Monterrey (Chapter 4) but in contrast with that observed in Bangalore (Chapter 3). The 1200 Wolf Street block captain reflected that the contest organizers:

> kind of downplayed the whole climate change and up-played the you are going to get a free roof coating kind of a thing and I know in getting people's signatures I definitely focused more on the personal gain of having energy efficiency upgrades in your house partly because I think a lot of people just weren't familiar with what any of it means or why that would be important to them.
>
> *(Interview 6, Resident, Philadelphia, October 2011)*

The challenge of facilitating ongoing awareness of energy usage and behaviour change also appeared to fall disproportionately on the block captain, highlighting the way the contest shifts the responsibility for governing climate change onto particular members of the community, by creating carbon-relevant subjects through the issue of energy efficiency. For instance, one resident was convinced that climate change was responsible for changing weather patterns in Philadelphia, and of the opinion that the City should be funding more grants for people to install solar panels. But another, though she was really happy with the Coolest Block contest, was concerned that they did not clean the roof as part of the works. She explained that the contractors had only spent a couple of hours at her place installing the white roof, and appeared to be unaware of who had organized and funded the contest, suggesting that the block captain would know that information (Interview 7, Resident, Philadelphia, October 2011). While the contest attempted to govern climate change through the community of the Block, it is clear that this is a partial process. By creating carbon-relevant subjects through the issue of energy efficiency, the contest actively devolved responsibility for climate change action to a network of civically-engaged subjects – the block captains – such that its climate change-related aims were more or less transparent, depending on how these block captains decided to sell the contest to their neighbours.

The partial means by which the contest has enabled retrofitting to become normalized within residential blocks raises questions about the long term impacts of the contest, highlighting the gap between what the experiment seeks to do in terms of repairing housing, intervening on rooftops and creating a legitimate mechanism for taking action on housing, and what it is *able* to do. The retrofits installed as part of the contest are not one-time solutions, but require maintenance over time. The cool roof coating, in particular, requires periodic re-coating to maintain its reflective characteristics. The respondent from the coating manufacturer Acrymax explained that the white reflective roof coating used on 1200 Wolf Street 'we would actually consider that about a five years system. Generally what they're going to do after five years is power wash, clean that off and do one refresher finish coat of everything' (Interview 8, Corporation, Media PA, October 2011), though he acknowledged that:

> we're a little pre-emptive in certain cases in saying you should recoat just to be on the safe side. You know we've seen people do that full system and not recoat it for close to seventeen years and it's still going strong.
>
> *(Interview 8, Corporation, Media PA, October 2011)*

This is particularly the case given that the Acrymax coatings used on their roofs are a premium product, targeted primarily at commercial buildings. As the representative from Acrymax observed, they do not routinely target homeowners in their marketing because homeowners typically are not willing to pay extra for their quality product, when similar-looking product at Lowes or Home Depot is sold for half the cost. Acrymax participated in the contest due to its existing relationships with Dow and the ECA, but perhaps also in the hope that the contest would stimulate neighbourhood-scale retrofits which would be a better fit for its business model than single-dwelling projects. A significant challenge could therefore emerge when this five-year maintenance interval is first reached. By that stage, the white roofs will become very much part of the infrastructural lattice of the block, but will also be less prominent in the consciousness of the residents, both because they are unseen and because of the time interval. Despite creating carbon subjects through the contest, the organizers cannot control the actions of these new subjects. Though the residents of 1200 Wolf Street are now carbon subjects, able to respond to climate change through their actions, they do not necessarily appreciate either their new power or the responsibility attached to it. Without any substantive investment (either financial or personal) in the new roofs, it is unclear whether residents will be willing to spend *their* money maintaining them. Moreover, in engaging with wider narratives of individual action to tackle climate change and personal benefit, the experiment deploys a particular form of subjectivity, anchored in personal gain, which does not speak directly to the collective endeavour that the experiment is, simultaneously, trying to promote.

Impacts and implications

The Coolest Block contest was assembled through the alignment of public–private interests, and was maintained to the extent that the winning block received the energy-efficiency retrofit upgrades promised. After the contest, public–private partnership models gained purchase as an effective means of achieving environmental and social goals. For Dow representatives, the Coolest Block contest demonstrated their leadership. For the City, the Coolest Block contest provided a model of neighbourhood retrofit that was put forward in subsequent applications to the Energy Efficiency Community Block Grants programme of the Department of Energy. Alignment was achieved both through leveraging the strengths of each partner and through an ethos of negotiation and dialogue in regular meetings, in which different forms of calculation were put forward with varied results.

The technical solutions being installed were less novel in comparison with the delivery of new institutional models. Indeed, some respondents, particularly those working for the City, showed a reluctance to describe the contest as 'experimental', not just because the notion of paying experiments with tax payers' money suggested carelessness, but also because the technologies implemented were not considered particularly new or revolutionary. For the coatings manufacturer, Acrymax retrofitting was not so much related to technical innovation as to the recognition of the availability of solutions. However, the experimental character of the Coolest Block contest is manifest in the application of such technologies to existing residential blocks at the neighbourhood scale. The Coolest Block contest mobilized climate change discourses in support of the renovation of the city's building stock. Location matters, not just because the Philadelphia setting provides an institutional context in which new public–private alliances can be forged, but also because the specific forms of materiality at work – from the city's planning to the architectural features of the blocks – determines the extent to which certain technologies succeed in embodying the twin purposes of addressing climate change and renovating the housing stock.

The Coolest Block contest may have also promoted technical innovation in the generation of a new architectural lexicon that can address the challenges of retrofitting and generate new ideas about how to achieve it. For example, the reconceptualization of the cool roof as an 'exoskeleton' of insulation, rather than as a barrier to weather, challenges conventional ideas of the housing block, specifically, the idea that insulation has to be done from the interior of the home. The notion of exoskeleton is particularly suited for an aging housing stock such as the one found in Philadelphia, in which roof cavities are frequently inaccessible or in which insulation batts are impossible to install because brittle electrical wiring and archaic wiring methods pose dangers both to installers and residents. In this context, technologies such as the cool roof constitute an innovation that may help circumvent the specific technological challenges of insulation in that particular location.

The contest also created a new means to engage with local residents in action to address climate change and reduce energy use in the city. In particular, the contest validated the neighbourhood scale as a valid scale of engagement for climate change governance in Philadelphia. This has led to the adoption of 'social norming' ideas in subsequent programmes, such as *EnergyWorks*, as a strategy to intervene in community networks through champions already embedded in these networks who can promote energy efficiency to other individuals. This is an approach suited to the tight-knit neighbourhoods that characterize many residential blocks in Philadelphia, often regarded as a 'city of neighbourhoods'. Those who sought to mobilize the community (e.g, ECA) found that 'social norming' made energy infrastructure visible, thus raising awareness among residents about the need to engage in energy-efficient practices. Simultaneously, the Coolest Bock contest contributes to a growing interest from different institutions to engage directly with communities at the neighbourhood level through the promotion of neighbourhood competitions, as neighbourhood rivalry is regarded as a dynamic factor that may increase motivation for weatherization work in adjacent communities.

The Coolest Block contest also aligned the city's emerging sustainability agenda and the imperative of addressing questions of economic redevelopment through the improvement of urban infrastructures. In a city such as Philadelphia, such questions are deeply tied to questions of inequality and exclusion, with the inner-urban population, predominantly non-white, concentrated in poor-quality housing and suffering greater levels of poverty. Retrofitting housing stock in the City has been driven by a concern to improve the living conditions and levels of fuel poverty in the city, as much as addressing issues of climate change. The organizers of the Coolest Block contest established poverty as one of the key challenges to both improving the quality of housing and improving Philadelphia's ability to adapt to a changing climate, whilst also reducing its contribution to greenhouse gas emissions. For the ECA, white roofs solved two problems at once. First, they provided a rationale for a badly-needed renewal of roofs in Philadelphia. Second, by acting as a form of 'external insulation', they would circumvent the challenges that hazards, such as old electrical wires, posed to cost-effective energy-efficiency retrofits, such as roof cavity insulation. The cool roofing and energy efficiency retrofits showcased in the Coolest Block were presented as affordable and, thus, as responding to the needs of low income housing.

However, despite the ECA's history in facilitating the weatherization of 'low income' homes, the Coolest Block contest did not really target low income blocks or people, and nor do the programmes that have followed the contest, such as *EnergyWorks*. *EnergyWorks* leverages resources from both public (ARRA, US$25 million) and private investment (US$25 million) to establish a revolving loan fund through which both expertise and resources are provided to residents interested in the programme. These incentives for energy efficiency upgrades, however, are available to homeowners who already have the ability to act and who can already access the loans to fund them (even if the interests are provided at very competitive rates). For example, the need to have a good credit score to access the funding is a clear limitation in a city in which loan rejections stand at 40 per cent among interested households. Overall the Coolest Block contest and the subsequent actions that have followed offer limited opportunities for those who might be most in need of retrofitting their houses. At the same time, by emphasizing the need to commercialize products (in this case roof coatings) and develop new markets, the Coolest Block contest clearly positions private capital as a valid beneficiary of climate change action. Thus, while the rhetoric of intervening in social housing to address climate change and energy poverty simultaneously is strong, the action space is limited to those who have already capacity to act and access to the necessary resources.

The Coolest Block contest emerged out of an initial alignment of interests, which resulted not just from deliberate efforts to engage, but also from the casual confluence of the conditions that facilitated such alignment. Yet, through this alignment, promoters became strategic in assembling the experiment and, particularly, in rendering it compelling to potential participants. Experiment maintenance and metabolic adjustment has fostered further circulations of meaning that have been used to justify a broader shift to public–private partnerships to enact climate governance, and an increasing use of community-based initiatives to create climate-relevant subjects in the city. But residents' subjectification, as far as this analysis shows, remains incomplete. Tensions remain between the focus on individual subjects, which is shown by the discourses of personal gain that promoted the experiment, and the community actions that were expected from residents collectively. These tensions had the effect of playing down the connection of the experiment with climate change within the contest, and thus enabled its presentation as a success. After the contest, however, these tensions have also exposed community interests which were overlooked in the experiment and how those very urban politics that seek to improve the living conditions of the poorer residents make them invisible.

References

Byrne, J., Hughes, K., Rickerson, W. and Kurdgelashvili, L. (2007) 'American policy conflict in the greenhouse: Divergent trends in federal, regional, state, and local green energy and climate change policy'. *Energy Policy*, 35(9): 4555–73.

City of Philadelphia (2009) *Greenworks Philadelphia, City of Philadelphia*, Mayor's Office of Sustainability. Available online at http://www.phila.gov/green/greenworks/2009-greenworks-report.html (accessed 30 June 2011).

——(2010a) *Mayor Nutter to celebrate with winners of Philly 'Coolest Block' contest.* Press Release, 18 June 2010. Available online at http://cityofphiladelphia.wordpress.com/2010/06/18/mayor-nutter-to-celebrate-with-winners-of-philly-%E2%80%9Ccoolest-block%E2%80%9D-contest/ (accessed 30 June 2011).

——(2010b) *Greenworks Philadelphia 2010 Progress Report.* City of Philadelphia, Mayor's Office of Sustainability.

——(2011) *Citywide Vision: Philadelphia2035*, Philadelphia: Philadelphia City Planning Commission (June 2011).

Pennsylvania Environmental Council (2008) *Building Green: Overcoming barriers in Philadelphia.* Pennsylvania Environmental Council Inc. Available online at http://www.pecpa.org/buildinggreen (accessed 30 June 2011).

6

FINANCING CLIMATE CHANGE EXPERIMENTS ACROSS CAPE TOWN'S HOUSING INFRASTRUCTURE

with Jonathan Silver

Introduction

Mamre is a peripheral neighbourhood in the northern boundaries of Cape Town that has provided a critical site to address issues of vulnerability and energy security together with climate change. With investment from the Danish International Development Agency (DANIDA), the City of Cape Town undertook a retrofitting project to install insulated ceilings in low-income houses in Mamre. The project aimed to combat problems of cold, condensation and ill-health experienced by those living in existing government-funded housing. This project has demonstrated the possibilities of international climate capital, while at the same time it has provided the municipality with space to experiment with new material configurations that are improving the quality of publicly-financed housing.

In this chapter, we focus on the making of the experiment, that is, how it was made in relation to dominant discourses of climate change, poverty and urban housing, and the forms of alignment and assembly that were required to bring the project into being. We find that the Mamre Ceiling Retrofit project is related to Cape Town's aspirations to become a national exemplar of green technologies and urban climate action. At the same time, climate finance has provided an opportunity to address the legacy of inadequate housing that intended to meet an urgent need in the immediate post-apartheid era. Housing provision over the last two decades has generated significant debate about the lack of quality of these new housing systems and the effects felt by low-income communities due to this underinvestment. Framing the issue of inadequate housing in climate change terms – both as a matter of vulnerability to existing and potential future climate impacts, and in terms of the significant drain on energy and household resources that providing thermal comfort within such houses entails – the City of Cape Town has been able to draw on additional resources, through the growing wealth of climate change finance, to address an ongoing urban challenge. In this case, a will to improve has been deployed through the development of mechanisms to access international carbon finance for local projects. The case also demonstrates the tensions that emerge as a result, such as the increased costs of delivering these levels of housing quality in parallel with addressing housing demand. Overall, the case demonstrates the importance of aligning multiple urban imperatives that move beyond a focus on climate change to incorporate a range of wider concerns that may provide a platform for social justice campaigners to demand higher quality housing for the urban poor, and a dramatic increase in climate change financing for low-income communities such as Mamre. The project is particularly significant in a context in which, though climate financing provides a platform for the municipality to rethink its housing policy outside current governing configurations, the progress to finance large-scale urban responses to climate change imperatives has been slow.

Poverty, development and climate change finance in Cape Town

The highly differentiated patterns of GHG emissions and vulnerability to the impacts of climate change are no more starkly apparent than in the context of urban South Africa. Cape Town has a Gini coefficient of 0.67 illustrating the internationally high degree of income inequality across the city (UN-Habitat 2008). At the same time, as a rapidly industrializing country, with a historic reliance on carbon-intensive coal-fired power generation (Bond and Erion 2009), South Africa has high total GHG emissions of 451,839 thousand metric tons of CO_2 per annum, which stands at 8.9 tons per capita (IEA 2011) and is ranked tenth globally. South Africa has come under increasing international pressure, along with Brazil, China and India, to address its contribution to climate change. At the Copenhagen Climate Summit in 2009, President Jacob Zuma pledged to reduce South Africa's GHG emissions by 34 per cent by 2020 and 44 per cent by 2025, contingent on the provision of financial support and technological transfers from those 'Annex 1' countries who have historically produced the highest levels of GHG emissions (NDRC 2013). In seeking to make significant inroads into its GHG emissions, South Africa has turned both to the development of new forms of energy economy, through the development of wind and solar energy, as well as to programmes of energy efficiency, both of which have partly been supported by new flows of climate finance. The World Bank (2010) estimates that climate change generates a need for capital investment into the global South of up to US$175 billion per annum for mitigation and up to US$100 billion per annum for adaptation. Current circulations of capital for climate change interventions remain far below these needs, with current estimates at about US$8 billion per annum for mitigation and US$1 billion per annum for adaptation (World Bank 2010), pointing both to the future possibilities in this arena, but also to the current lack of action to address climate change globally which will have profound consequences for vulnerable communities. In the remainder of this section, we first introduce the urban development context of Cape Town, and in particular the ways in which the acute need for housing provision has been addressed to date, before considering how the City of Cape Town has sought to position itself in relation to the emerging drive to access climate finance as a means of addressing development challenges.

Urban geographies of inequality and housing provision

The City of Cape Town, alongside the national and provincial governments, has invested billions of Rand to address poverty and inequality over the past two decades (Parnell *et al.* 2005). Despite this investment, this is a city in which socio-economic inequality is evident, stark and ever-present, reflecting the segregated, racialized landscapes that constitute its geography. As Cape Town approaches two decades since the end of apartheid, the municipality remains committed to developmental policies, including the large-scale transformation of housing and infrastructures. These post-apartheid policies have brought significant public investment through the Reconstruction and Development Program (RDP), resulting in housing units being built each year since 1994 (Mongwe 2011). These flows of investment into housing infrastructure take place in a context of widespread poverty and socio-environmental injustice, with up to 400,000 people waiting for new housing and many thousands living in informal settlements lacking basic services, generating ongoing anger and disillusionment (Mongwe 2011). In this context, the delivery of RDP housing provides the key means for the City of Cape Town to empower the city's urban poor and has become a mainstay of municipal policy, in which the provision of housing and creation of home ownership operates as the de facto poverty alleviation strategy (Lemanski 2011). Yet for many of the urban poor, acquiring an RDP house does not easily translate into direct improvements in livelihoods, health and other indicators of socio-economic status. Lodge (2003), for example, shows that around 30 per cent of new RDP houses fail to comply with building regulations, which, together with the low overall build quality, leaves many households with a series of connected vulnerabilities and little likelihood of having the ability to generate or receive further investment in the house. The estimated cost of repairing the low-quality RDP houses across South Africa is around R58 billion (around £450 million) (News24 2011).

Moreover, the developmental aspirations of the state, focused on delivering post-apartheid promises to its residents, often appear to conflict with the broader macro-economic commitment to neoliberalism which has increasingly shaped urban governance (McDonald 2008). In Cape Town, such struggles are also manifest in the city's politics. The city is governed by the Democratic Alliance, the main opposition party to the national ruling African National Congress party in South Africa, which uses Cape Town as a showcase for its national political agenda. Issues of service delivery, poverty, economic development and competence constitute the arenas of political confrontation. The nature of Cape Town's infrastructure politics is also shaped by its status as a well-resourced South African municipality together with the fact that the municipal government derives a key revenue stream from its role as the electricity distributor in many parts of the city. The capacity to experiment with housing and service delivery projects, as well as to link up with international flows of finance, demonstrate both the relatively well-resourced nature of infrastructure provision and its contested status within the city.

The brave new world of climate finance in South Africa and Cape Town

South African cities have developed various responses to climate change imperatives for over a decade, linking these logics into the wider discourses of sustainable development that have particularly accelerated into issues of energy efficiency and mitigation since the 2008 energy crisis. Claims by municipalities, including eThekwini and Cape Town, to be the leading South Africa city in responding to climate change have helped to shift debates and policy development, yet these efforts are stymied by the failure to decarbonize energy supply, transform urban planning beyond apartheid geographies or low-density landscapes, and the lack of financial resources to make significant progress in addressing climate change imperatives. Municipalities are increasingly linking into global networks such as C40 cities, and working alongside other actors such as ICLEI to develop further responses.

Across Cape Town, various groups and organizations are involved in developing new technologies, infrastructures and policies to address climate change imperatives. Recognition of Cape Town as a perceived hub for innovation and technological development has come in prizes such as the South African Climate Change Leaders Awards, the World Design Capital 2014 and Role Model City by the United Nations Strategy for Disaster Risk Reduction. The City of Cape Town pioneered the development of an *Energy and Climate Change Strategy* (City of Cape Town 2006) in Africa. This was followed up with the *Energy and Climate Change Action Plan* (City of Cape Town 2010). Municipal planning is coupled with local initiatives which enrol civil society actors and publics, such as the energy efficiency SMART Living campaign, the public transport MyCiti Bus Rapid Transport system and the small scale Darling Windfarm. The City of Cape Town presents the city as a 'brand' that distinguishes the city as a hub for best practice and innovation through a vision of 'a sustainable, world-class African city' and 'a leading city in South Africa, as well as in the continent' (City of Cape Town 2006: 5). New coalitions of urban intermediaries, incorporating a range of state and non-state actors are coalescing (if often for very different reasons) around the notion of the sustainable (African) city, and becoming involved in climate change action through multiple and often competing logics from economic development, capital accumulation and city branding through to socio-environmental justice, a greener Cape Town and a safe city.

It is in this context that the City of Cape Town began considering ways to access climate change financing to improve urban infrastructure across the city, begin adaptation and mitigation, and create a low carbon vision of the urban future. Beyond the provision of the Kyoto Protocol for those Annex 1 parties to the UN Framework Convention on Climate Change to invest in carbon reduction projects in the Global South through the Clean Development Mechanism (CDM), climate change financing is instigated by a plethora of bilateral, multilateral and non-governmental donors with competing strategies and policies (Bulkeley 2013). It also depends on changing fiscal and political conditions across donor countries. Similar to financial aid flows (Beltran 2012), climate change finance may be more responsive to the priorities of donor countries (such as opening new markets) than the needs of recipients (Bond 2011). Moreover, climate finance depends upon institutional arrangements which require a great deal of expertise related to, for example, climate change science, low carbon and resilient technologies, and international financing. Amassing the expertise to access and secure carbon finance is often beyond the capabilities of municipal governments which need to work with a shifting series of guidelines, rules and assessments. According to the OECD (Clapp *et al.* 2010: 11), the predominant flows of carbon finance are through the 'compliance market' (the CDM and Joint Implementation) and 'urban mitigation projects represent less than 10 per cent of all projects in the compliance market today and are concentrated in few sectors (waste management, energy efficiency, and energy distribution networks)'. Furthermore, Africa belongs to the periphery of the global financial system and climate finance is not an exception, in spite of being one of the most vulnerable regions to the effects of climate change (Toumlin 2009). Finance also requires a favorable regulatory environment, financial creditability and leverage of existing resources at the local level, suggesting climate finance is mostly available for well-resourced and 'on message' municipalities. This means that even relatively well-resourced African cities, such as Cape Town, find it challenging to access climate finance through the compliance and voluntary carbon markets, as flows of aid, or through the growing forms of private investment that appear to be occurring in low carbon urban developments in cities in Asia (Bulkeley and Castán Broto 2013).

In order to overcome such challenges, multilateral institutions, such as the Cities Development Initiative for Asia (CDIA), argue that municipalities should use technical assistance and create partnerships with the private sector, NGOs and other non-state actors to access climate finance (Beltran 2012). At the same time, the resources, knowledges, institutional and public support generated by initial climate change responses, such as those which the City of Cape Town have undertaken, offer a context in which climate change experiments may emerge, attracting donor attention and opening up financing pathways for future experimentation and broader societal change. Such experiments may help South African cities such as Cape Town to access flows of investment by demonstrating the possibilities for intervention provided by climate finance and how its institutional complexity can be managed.

The Mamre Ceiling Retrofit experiment: securing climate finance for social justice?

Situated on the northern boundary of the city, Mamre is a township of 10,000 people and characterized as low income with an unemployment rate of over 28 per cent (City of Cape Town 2013). The area has been part of an often fraught process of state investment to provide 500 RDP housing units to many deprived households in the area (Davy 2006). These houses are of basic construction, consisting of walls, roofs and service points, and were financed and constructed as part of the more than three million publicly-funded houses built in South Africa since apartheid (Figure 6.1). Although publicly funded and built by the state, ownership of RDP housing is transferred from the state to the household upon completion. Challenges have emerged during the development of this form of housing provision, including issues of environmental sustainability and the potential of the housing to address health and livelihood issues for the poor (Goebel 2007). These issues have been particularly acute in Mamre, which is located in the Southern Cape Condensation Problem Area (SCCPA) and, as a result, vulnerable to conditions of damp, cold, wind and rain that characterize its winter climate. These meteorological dynamics are particularly challenging for low-income households, who often lack the finances to sustain flows of energy to keep their homes warm and dry. Since 2003, the Western Cape region has received extra subsidy from national government for RDP house construction to ensure that insulated ceilings – not previously part of the design of RDP housing – are built into any new development, to provide thermal protection against the climate. However, this subsidy does not reach around 40,000 RDP households in the Western Cape region built before 2003. These dwellings, which

FIGURE 6.1 A house with a retrofitted ceiling. Photo © J. Silver, used with permission.

have low thermal protection, generate numerous health problems for households demonstrated, for example, in the high frequency of pathogenic metabolisms of pneumonia, tuberculosis (TB) and flu in the region.

One possible strategy for dealing with the legacy of inadequate housing provision and the challenges they create is to retrofit existing RDP housing, to improve thermal and energy efficiencies and better address socio-environmental conditions for households. So, in 2010, the community of Mamre became one of the first areas in Cape Town to receive investment through international climate change financing. The investment financed a retrofitting project which aimed to provide insulated ceilings for 240 of the RDP houses. It required an alignment between the interests of (international) climate change finance and the municipality and community, as well as the development of methods to account for the reduction in GHG emissions that such an intervention could produce. The project was called the Mamre Ceiling Retrofit project and led by the Sustainable Livelihoods section of the Environmental Resource Management Department of the City of Cape Town (alongside the Housing Department, Spatial Planning Department and other divisions within the municipality), and financed by the DANIDA Urban Fund through its *Promoting Resilience of at Risk Communities in Climate Change* programme, with the support of participating households. Whilst not a new technology, this mode of retrofitting represents an innovative attempt to reduce energy poverty, increase climate resilience and provide livelihood improvements through accessing a new financial mechanism and through bringing together a novel range of partners to address the problem of housing quality. Projects like the one in Mamre show how international funding seeks to find sites for intervention through mobilizing climate change discourses. In the rest of this section, we examine how the project was made, the ways in which it became maintained as adjustments to the metabolic flows of energy and pathogens took place within the community, and how it was lived through the processes of engaging the community in the everyday practices and economies of ceiling retrofits.

Making: the growing significance of international climate change financing in Cape Town

The first initiative in Cape Town to experiment with the possibilities of climate change finance for addressing the broader housing and infrastructure challenges facing the city was not the Mamre Ceiling Retrofit project, but rather the Kuyasa Clean Development Mechanism (CDM) project in Cape Town's largest township of Khayelitsha, which was designed to retrofit houses with solar hot water (SHW) heaters, insulated ceilings and other energy savings measures, in order to demonstrate the potential for climate change finance to deliver wider sustainability benefits for poor urban communities in South Africa. The Kuyasa project was a vital part of the making of the Mamre Ceiling Retrofit project, because it brought relevant actors into alignment and began to render climate change governable through an assemblage of international and local actors facilitated through carbon finance. Led by the NGO SouthSouthNorth and the City of Cape Town, the project undertook a three year design phase before being registered as a CDM project in 2005. An initial 'ten pilot houses were adapted in 2005, providing the practical data for an efficiency measuring system' (Kuyasa CDM 2013) such that the carbon savings that could be accrued from a package of energy efficiency and solar energy measures could be calculated and their benefits assessed in order to demonstrate that such an intervention could meet the requirements for accessing CDM finance.

The project has generated considerable interest, given that 'Kuyasa is South Africa's first internationally registered Clean Development Mechanism (CDM) project under the Kyoto Protocol on climate change and was the first Gold Standard Project to be registered in the world' (Kuyasa CDM 2013). Following its registration as a CDM project, the financial support for completing the initial phase of the project was provided by 'the Provincial Government of the Western Cape (PGWC) and ICLEI (~R4.7 million) and the National Department of Environment and Tourism (DEAT) through its Poverty Alleviation Funding of approximately R25 million' (City of Cape Town 2013). Between 2008 and 2010, this funding supported the retrofit of over 2,300 houses, implemented by the South African Export Development Fund. The project has generated income, through the sale of carbon off-set credits, via the CDM, which is used for ongoing maintenance and the future expansion of the scheme, although this remains vulnerable to the fluctuating market for carbon credits (Wlokas and Corbera 2009).

The Kuyasa CDM project formed an important focus point for the City of Cape Town and its efforts to marry a concern to place the city at the heart of new initiatives to develop green technologies and urban sustainability, on the one hand, and to address the long-standing challenges of development and poverty on the other. In so doing, the project provided an important means through which the problem of climate change in the city came to be understood as one of both addressing climate vulnerability – the cold and damp conditions that much of the population experiences – and energy security through the arena of addressing the inadequacies of existing housing. Regarded in these terms, the key technologies deployed through Kuyasa became central to the policy agenda of the City of Cape Town – solar hot water and ceiling insulation – laying the ground for such projects as the Mamre Ceiling Retrofit project. These technologies have been promoted through a range of different actions. These include a partnership between the municipality and ESKOM to deliver over 100,000 free solar water heaters, the establishment of an accreditation scheme for this technology targeted on increasing usage by middle- and high-income households and guidance on the best forms of insulation materials. Furthermore, with its emphasis on the need to calculate not only the carbon savings incurred – a critical aspect for accessing carbon finance – but also the wider livelihood benefits, the project established a set of parameters, tools and techniques through which the assessment of the potential impacts of such interventions could be accounted for and legitimized. The Kuyasa CDM project was therefore critical in problematizing climate change in relation to low-income housing, creating knowledge about the forms of calculation required, and demonstrating the means through which additional resources could be secured towards these ends.

As a pilot project, conducted with a range of partners and only partially framed in housing terms, the Kuyasa CDM project enabled the municipality to experiment outside its normal service delivery framework, which has been guided by the logic of constructing as many housing units as possible, to consider alternative goals for the provision of housing. The Mamre Ceiling Retrofit project emerged within this milieu, but departed in significant ways from the Kuyasa CDM project. Rather than seeking to pilot and invest in a project in order to attract carbon finance, in the case of Mamre, funding was secured through a bilateral donor, and the municipality then determined the means through which it should be invested. While the Kuyasa CDM project had opened up the potential for accessing finance through the compliance market, it had also demonstrated the significant challenges and barriers to doing so – not least of which were issues of calculating and verifying the savings incurred against predicted baselines, and the upfront costs of doing so for multiple small projects (Clapp *et al.* 2010; Kuyasa CDM 2013). Instead, the City of Cape Town focused on developing ways to incorporate such interventions within its climate change adaptation and mitigation actions. Municipal policy makers secured the financing of R1.9 million (US$300,000) from DANIDA based on a wider programme of donor aid to the city for climate change adaptation and mitigation. At the Environment Resource Management department of the City of Cape Town, policymakers from the Sustainable Livelihoods section were given a fixed amount of funding with relative freedom to consider the best intervention in relation to green energy or energy efficiency. Framed as an issue of improving the housing stock in the city and building on the findings from the Kuyasa CDM project, policymakers promoted ceiling retrofit, because of the possibilities it offered for improving the life of a large sector of the population with relatively low per capita investment.

Once the notion of ceiling retrofit had gained traction, the community of Mamre was selected for the project and allocated the financing because of its relatively small size, providing the opportunity for the City of Cape Town to undertake the retrofit of a significant part of the RDP houses in the neighbourhood (nearing 50 per cent) with the finance available. The combination of DANIDA's funding priorities with the intervention of policy makers demonstrates the role that particular urban actors can play in championing certain ideas, promoting particular technologies and deciding on the location of experiments. The retrofit of the houses was relatively straightforward, reflecting the established nature of ceiling insulation as a practice. The installation provided a ceiling and, by using 50mm of glass wool or polyester insulation for extra thermal protection, provides a relatively invisible addition to the households of Mamre (Figure 6.2). The insulated ceilings were fitted by a team of locally-trained workers during 2010, with priority for participation in the project given to certain vulnerable groups, such as households with disabled or elderly members and child-headed families. Throughout, the community of Mamre was involved in the selection process, delivery and the subsequent monitoring of the project.

The Mamre Ceiling Retrofit project emerges therefore at the confluence of a broader international move towards seeking new ways of enabling the uptake of climate change mitigation and adaptation across the global South, the interest of

FIGURE 6.2 A street in Mamre. Photo © J. Silver, used with permission.

financial institutions and donor organizations in climate change finance as both a new economic arena and a means through which to enact broader programmes for sustainable development, and the particular circumstances and pressures of addressing issues of vulnerability, energy security, housing quality and climate change in the City of Cape Town. Through aligning the existing problematic of inadequate housing provision with the newly problematized spectre of reducing GHG emissions in cities in the global South, the municipality created a space for intervention through which new flows of finance could be mobilized. Calculating such mechanisms of mobilization is a key aspect of the experiment. In Mamre, the alignment was forged gradually in the decade leading up to the intervention, through the partnership established with SouthSouthNorth, the Kuyasa CDM pilot project, and the growing interest of national government and international organizations in the potential for addressing energy efficiency in South Africa's urban housing provision. By insisting on the wider, sustainable development benefits of such interventions, both Kuyasa and Mamre have served to make such interventions compelling in a context where the challenges of providing adequate shelter, livelihoods and well-being for large numbers of the city's residents are urgent and pressing.

Maintaining: adjusting the metabolic flows of energy and disease

The installation of new ceilings and installation within the RDP houses at Mamre was undertaken as a one off venture, a simple technical intervention that would require no ongoing upkeep by the municipality or by the residents involved. Once installed, however, the insulated ceilings became caught up within various metabolic circulations that operate through and serve to sustain particular housing configurations and household dynamics, serving to partially disrupt and reconfigure their operation in the process through which the experiment became embedded within the city. Research, including a 150 household survey and 25 interviews undertaken with members of the community, found that the insulated ceilings served to entwine households with the daily operation of the experiment through three key metabolic circulations: energy, finance, and disease (Silver *et al.* 2011).

Sustainable Energy Africa (2010) estimate that insulated ceiling installation can improve the thermal efficiency of RDP housing by up to six times (see Figure 6.3). The effect for the majority of Mamre's households has been to lower the energy required to heat a domestic space for a period of time. With less electricity needed to heat the home, a series of metabolic adjustments have been made to the energy flows which sustained housing provision, in turn reconfiguring these circulations within the urban context. First, the intervention has had an effect on the use of non-electrical heating sources, such as wood-fuelled fires and kerosene for heating, which have caused extensive incidents of death, injuries and damage across the city (Pharoah 2009). The research found a significant reduction in households using fire to heat houses from 52 respondent households before the intervention to 25 afterward (Silver et al. 2011), in turn leading to significant reductions in the circulation of hazardous toxins inside the houses, as well as to reduced markets for these forms of energy in the neighbourhood. Second, while for many households (56 per cent)

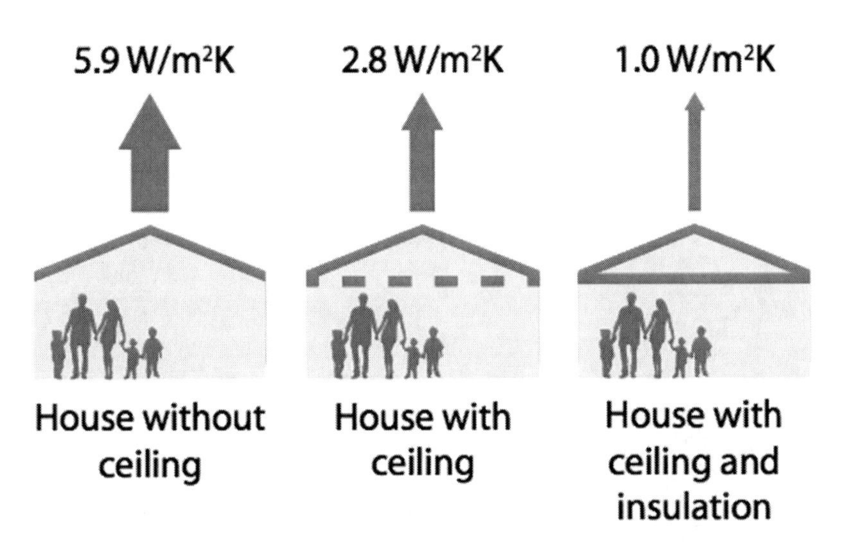

FIGURE 6.3 Analysis of thermal efficiencies in RDP housing. Reproduced from Sustainable Energy Africa (2010). Design: R. Hufton, used with permission.

the overall amount of energy used in the household has not been reduced, the ways in which energy is used have changed (Silver et al. 2011). For some, the intervention has allowed homes to stay warm for longer due to the increased thermal efficiency provided by the insulated ceilings. As one community member commented: 'We are using the same amount of electricity as before but [it] lasts longer so can help the house stay warm for longer' (Interview 18, Resident, Cape Town, May 2011). For the majority of households that are using less energy for heating, the household survey showed that for many (around 65 per cent in that category) the energy is simply used for other appliances requiring electricity. As such, households are choosing the 'take back' effect, in which the financial saving is reallocated within the energy budget to meet the repressed (energy) demand of the household.

As this evidence suggests, the metabolic circulation of energy through the city is closely related to that of the flow of household finance. Whereas for the majority of households at Mamre, flows of finance have been redirected through the reconfiguration of energy services, for just over one third of those who participated (35 per cent) the energy savings generated by the insulated ceilings are used to support other areas of the household budget. As one participant explained: 'We don't use as much electric now after ceiling has been installed so save money, which is now spent on food for the children' (Interview 8, Resident, Cape Town, May 2011). However, the ceiling retrofit has not been sufficient to completely adjust metabolic conditions in relation to required flows of finance to provide energy services, with nearly a third of households (31 per cent) involved in the ceiling intervention continuing to struggle with the costs of energy in relation to current energy needs (Silver *et al.* 2011), suggesting that the improvements to housing quality afforded by these interventions have been only partially able to readjust and reconfigure RDP housing in relation to the climatic and economic conditions of Mamre.

At the same time, the research found that the interventions had served to disrupt the circulation of other forms of metabolism and to sustain improvements in health and well-being. Some 67 per cent of respondents involved in the experiment said they had noticed an improvement in family health since the ceiling installation (Silver *et al.* 2011). For instance, the research recorded a significant reduction in the frequency of asthma and other breathing difficulties, reducing from 24 households to six, with a similar pattern being recorded for cold and flu symptoms from 49 households to seven households. These examples show how altering the flows of energy into the house can also shift a range of other socio-natural circulations that affect health. TB is one such example of the linkages between retrofitting and health improvements. In South Africa there are around 419 cases per 100,000 people, one of the highest TB rates in the world (Fourie 2011). In Mamre this has historically been a problem, reinforced by the poor housing quality of the RDP houses. Before the ceiling retrofitting, poor air quality and the presence of damp, smoke and mould influenced the spread and impact of TB. The household survey shows that TB has been reduced over the period in which the ceilings were installed. Households with insulated ceilings have experienced a reduction in the frequency of TB through increased thermal protection, with households suffering TB reduced from 31 cases the winter before to five cases the winter after installation (Silver *et al.* 2011). The ability of households to sustain their livelihoods also depends on good health. For many of Mamre's residents their livelihoods are precarious and prone to disruption from ill health, in turn making life more difficult for poor households to generate both basic needs and opportunities to improve socio-economic conditions. By reworking the circulation of heat, condensation, mould and pathogens, ceiling insulation goes some way in mitigating this health-influenced precariousness.

In the case of the Mamre Ceiling Retrofit project, therefore, we find that the process of maintenance is predominantly realized through the ways in which the installation redirects and reconfigures flows of energy, finance and health within the neighbourhood. In so doing, the experiment becomes embedded within this urban milieu, whilst at the same time its benefits (and limitations) are circulated through flows of economic activity, energy markets, education, bacteria and clinical interventions, all the while adjusting the wider urban political ecology to the presence of the experiment itself.

Delivering development and reconfiguring livelihoods through living the experiment

Through improving the thermal efficiencies in Mamre's RDP houses, the City of Cape Town and the partners involved in this intervention seek to improve the lives of its residents in a fundamental way. However, how such 'improvements' play out is only revealed as time passes and the innovation becomes embedded in the spaces of daily experience, as the project comes to be embodied and lived through the daily practices and livelihoods of the residents themselves. A critical means through which the experiment has come to be lived is through the engagement of the community in Mamre throughout the course of its initiation, delivery and assessment. In so doing, particular forms of community and household subject were envisaged, who would benefit from the ways in which the installation of insulated ceilings would offer an improvement on the existing experience of living in RDP homes. As in Philadelphia (also a low carbon retrofit project), rather than stressing the low carbon benefits of the project, either in terms of the ways in which it had been realized through donor-led climate change finance or the imperatives for residents to become low carbon citizens, residents were engaged with the notions of a warmer, more comfortable home, that would provide higher levels of energy security and affordability. Here, the emphasis was on the wider benefits of the installation, and the ways in which it would enhance quality of life through providing dignified housing.

The type and location of the experiment was shaped within the municipality, leaving the Mamre community little flexibility in influencing the making of the project. There was however significant community involvement in the delivery of the insulated ceilings as both the City of Cape Town and the NGO ICLEI Africa worked with residents to stress the connected benefits of the experiment, mobilizing discourses of improvement, dignified housing and climate resilience in the communications with households. A key part of this communication concerned the acknowledgement of the energy poverty issues of households in Mamre and recognizing the need to support poor communities beyond basic housing and service delivery. The process of community engagement included identifying the suitable households for intervention, training community research teams for monitoring and evaluation, and employing and training local labour to deliver the experiment.

The criteria for the selection in the project included: only original owners that were still residing in their RDP house (as significant number are often sold on); households that have not made significant alterations (including insulation) since taking ownership; and households that had not received free geysers from the Provincial Government. Preference was given to certain vulnerable groups, such as disabled members, child-head families and elderly households. For those households excluded from the process, there was anger about being left out of the process, particularly from those who had invested in housing improvement and were thus excluded from this project. As one such research participant commented, the process: 'could have been a fairer process by handling us the same, people don't know my household circumstances, how could they?' (Interview 9, Resident, Cape Town, May 2011). Overall the project was welcomed by the community with little evidence of any resistance or contestation of the experiment, and with 90 per cent of the survey respondents satisfied with the insulated ceilings. Perhaps the only area of contention was around the limited nature of the experiment, which did not deliver such improvement to all households in the community.

A second essential element to the project and its resonance within the community has been the emphasis on the direct economic benefits that would be realized through the implementation of the project. The public tender that was issued for the conduct of the project specified the use of local labour, and the research found that this had generated substantial (if temporary) benefits for twenty households in the neighbourhood:

> We made sure they use local labour with four teams of five people that were employed. Some of these people are now employed in other work whereas before they knew nothing now they work in private installations in other areas so good long term benefit. Currently there are eight guys who are now competent to do the job and can go out on their own and do the job. They could help the CCT [City of Cape Town] in other places, also they are learning thatching skills so expanding their skills. This is regular work, now earning R750 a week.
>
> *(Interview 13, Resident, Cape Town, May 2011)*

As this respondent makes clear, one of the outcomes of the project has been the training and long term employment of eight of these workers in related projects, accruing an estimated R250K in earnings to families in the neighbourhood (Silver *et al.* 2011). Central to the operation of this project, therefore, has been the development of new forms of subject capable of working on the (low carbon) improvement of housing in the city, and recirculating the benefits accruing from climate change finance through the local economy. At the same time, the project has also been taken up and embodied in the livelihood practices of the community through its effect on reducing ill-health. The research found that some 246 sick days had been saved over the year following the intervention, enabling households to sustain their livelihood activities, such that 82 of the 140 households avoided the costs associated with taking time off work due to illness. In this manner, the intervention has also served to indirectly redeploy climate finance towards supporting the economic activities of low-income families.

In the case of Mamre, the focus on the deployment of carbon finance has turned attention away from an overall emphasis on the residents within the experiment as the subjects towards which authority has to be conducted (as it is the case in other examples, such as T-Zed in Bangalore). Instead, here, there is an emphasis on the control of the flows of materials, skills and materials, translating the process of carbon subjectification from the citizens to those who make the experiment possible. As agents of urban change, a wide range of actors other than those who inhabit it influence the constitution of the development and how it is lived. Living does not just pertain to the activity of dwelling, but also it pertains to how the experiment comes to be lived more broadly, through the city, integrated in the collective imagination of the future as an engine of the will to improve. The constitution of subjects does not just occur in the daily life of those who we imagine inhabiting experiments, but also in the daily life of those who make them possible. In terms of the residents of Mamre, however, only a subset have been able to take advantage of the livelihoods provided by work in the retrofit sector opened up by the experiment.

Impacts and implications

The Mamre Ceiling Retrofit project provides an example of a climate change experiment that relies on the installation of basic technology, in this case insulated ceilings, together with innovative institutional and financial arrangements as a means through which to align responses to climate change with long-standing urban challenges. As such, this experiment is similar to the Coolest Block contest in Philadelphia, despite the very different urban contexts. The emergence of Mamre as a climate change experiment can be partly explained by the ability of the city's policy makers to align such interventions both with the climate change agendas of urban, national and international interests, as well as with existing policy goals and logics, particularly those of housing improvement. This problematization of responding to climate change as requiring the improvement of housing infrastructures in the city was achieved through advancing the twin logics of the vulnerability of existing housing provision to climatic change and the impact on poverty of the limited energy security such housing afforded. Aligning climate change with poor housing in this manner enabled simple technologies, such as retrofitting ceilings (in the case of Mamre) and solar hot water systems (in Kuyasa), to be regarded in a new light: as interventions that could at once enable the improvement of housing and provide a measurable impact on overall levels of energy consumption, yielding financial savings and wider community benefits.

Such logics were important both in addressing the priorities of DANIDA as a donor and the need for the City of Cape Town to consider how to access larger amounts of climate financing, enabling a range of actors to align themselves around such interventions. By offering a cost-effective technology that could produce measurable financial and health benefits, the Mamre Ceiling Retrofit project was also able to align concerns about the economic status of RDP households and wider economic development imperatives mediating urban policy with the opportunities afforded by the 'green' economic development logic installed through ceiling retrofit. The investment in energy and housing networks at the household level generates both direct economic impact (jobs, training) and longer-term considerations (such as improved health, lower energy bills). Finally, the experiment was also positioned as a potential way of addressing wider socio-economic conditions in Cape Town through reconfiguring existing RDP housing to improve its role in support of the city's urban poor. By acting as a conduit through which multiple urban problems could be aligned, the experiment thus acted to re-imagine the house as a site of environmental (climate change), and social (energy security, structural quality of housing) benefits.

In creating this form of alignment and convergence between the problem of climate change and the need to improve housing in the city, the Mamre experiment was dependent on the growing interest in climate change finance amongst donor organizations and government agencies. The global flows of climate finance and its circulation into infrastructures of the global South provide a range of opportunities for municipalities such as Cape Town to reinforce and expand development discourses, actions and logics through the metabolic adjustment of circulations of energy into low-income households. Yet the complicated, shifting and contradictory procedures and pathways to access financing often remain inaccessible to sub-national actors, meaning municipalities such as the City of Cape Town and their partners must continually experiment with ways to connect to such funding streams, develop innovative verification methodologies and donor priorities, and ensure that their initiatives generate co-benefits to address other municipal priorities. As such, whilst experiments such as Mamre that access international climate change finance remain few and far between, particularly in Africa, they serve to demonstrate the availability of both new revenue streams for climate change related interventions, and the existence of new techniques and tactics to secure this finance. This shapes the way in which climate change interventions come to be imagined and enacted in the city. In turn, the momentum from experiments such as Mamre and Kuyasa have enabled the City of Cape Town to develop particular representations of their achievements and advantages which mediate their relationship with donors, consultants and NGOs, in turn attracting further potential global flows of climate, carbon and aid financing.

Yet, at the same time, the momentum gained from such experimentation remains fragile and open to contestation. On the one hand, by enabling a different perspective on the provision of housing in the city – housing that not only meets the basic needs of the poor but offers greater levels of protection and security, whilst enabling more comfortable and affordable lives – the Mamre experiment challenges existing priorities and suggests new ways in which housing and energy provision can be financed. The experiment not only shows that the quality of the housing can be increased, but also that this can be achieved with a relatively small investment and through an incremental improvement in the materialities of RDP houses. Because of the role of the experiment in improving the quality of the RDP housing in Mamre, it may be poised to generate debate around the model of housing delivery practiced by the municipality. Historically, the assumption has been that basic housing is all that can be afforded in the context of significant need and limited finance. However, the Mamre project, by demonstrating the low costs involved and the wider economic benefits accrued, challenges such assumptions and points to alternative pathways for housing provision. However, if this finance is redirected away from the immediate task of housing people in desperate need, such financing arrangements are likely to remain deeply contested. Such considerations generate the need for policymakers to create further alignments between basic housing needs, improved housing construction, climate resilience and energy poverty and create significant impacts from these new streams of funding.

Overall, we find that the case demonstrates how experiments can alter the socio-technical and political-ecological relations in the city through creating space for debate and evaluation about the objectives of municipalities and other actors who are seeking to address problems of urban development. At the same time, we find that in cities in the global South, where climate change imperatives encounter development demands, emerging climate change adaptation and mitigation measures need to be aligned with existing policy imperatives, both because this is the only way to legitimate these interventions on the ground and because aligning objectives is a measure to ensure their sustainability in the long term (Satterthwaithe *et al.* 2007). The Mamre Ceiling Retrofit project shows that an intervention centred around logics of housing improvement can also incorporate a range of other imperatives, including mitigation and adaptation, energy poverty and security, economic and livelihood considerations and health. Across South African cities, the necessity of investment fulfilling multiple objectives is paramount as municipalities are forced to 'do more with less'. With large amounts of capital already circulating globally and significant increases in climate change financing predicted over the next few decades, the open question is whether the Mamre case signals the beginning of a new wave of investment targeted at tackling Cape Town's deep-seated problems, addressing issues of social and environmental injustice through shifting metabolic conditions and opening out an idea of the metabolic commons, something that should be welcomed in a city that remains characterized by widespread poverty and inequality. Yet such incremental change masks the widespread transformation of urban landscapes required to address climate change imperatives. If such new investment is to dramatically shift the terms of debate around social and environmental justice, then it is vital that wider coalitions of urban intermediaries are involved in debating, deciding and delivering such experiments.

References

Beltran, P. (2012) *International financing options for city climate change interventions: an introductory guide*. Manila: Cities Development Initiative for Asia. Available online at http://www.cdia.asia/wp-content/uploads/International-Financing-Options-for-City-Climate-Change-Interventions1.pdf (accessed 30 June 2014).

Bond, P. (ed.) (2011) *Durban's climate gamble: Trading carbon, betting the earth*. Braamfontein: Unisa Press.

Bond, P. and Erion, G. (2009) 'South African carbon trading: A counterproductive climate change strategy'. In McDonald, D. (ed.) *Electric Capitalism Recolonising Africa on the Power Grid*. London: Earthscan, pp. 338–58.

Bulkeley, H. (2013) *Cities and Climate Change*. New York: Routledge.

Bulkeley, H. and Castán Broto, V. (2013) 'Government by experiment? Global cities and the governing of climate change'. *Transactions of the Institute of British Geographers*, 38: 361–75.

City of Cape Town (2006) *Energy and Climate Change Strategy*. Available online at http://www.capetown.gov.za/en/EnvironmentalResourceManagement/publications/Documents/Energy_+_Climate_Change_Strategy_2_-_10_2007_301020079335_465.pdf. (accessed 12 March 2010).

——(2010) *Energy and Climate Change Action Plan*. Available online at http://www.capetown.gov.za/en/EnvironmentalResourceManagement/publications/Documents/Energy+CC_Action_Plan_project_info_2010–08.pdf (accessed 12 March 2010).

——(2013) *Demographic Profile of Ward 029*. Available online at https://www.capetown.gov.za/en/stats/Documents/2011%20Census/Wards2011_Census_CT_Ward_029_Profile.pdf (accessed 11 March 2013).

Clapp, C., Leseur, A., Sartor, O., Briner, G. and Corfee-Morlot, J. (2010) *Cities and Carbon Market Finance: Taking stock of cities, experience with Clean Development Mechanism (CDM) and Joint Implementation (JI)*. OECD Environmental Working Paper No. 29. OECD Publishing.

Davy, J. (2006) *Assessing Public Participation Strategies in Low-income Housing: The Mamre Housing Project*. MA Public and Development Management Thesis, University of Stellenbosch. Available online at http://scholar.sun.ac.za/handle/10019.1/1851. (accessed 13 April 2011).

Fourie, B. (2011) *The Burden of Tuberculosis in South Africa*. SA Healthinfo. Available online at http://www.sahealthinfo.org/tb/tbburden.htm (accessed 1 November 2013).

Goebel, A. (2007) 'Sustainable urban development? Low-cost housing challenges in South Africa'. *Habitat International*, 31: 291–302.

IEA (International Energy Agency) (2012) *CO_2 Emissions from Fuel Combustion*. Paris: IEA.

Kuyasa CDM (2013) *Kuyasa CDM Project*. Available online at http://www.kuyasacdm.co.za/ (accessed 1 November 2013).

Lemanski, C. (2011) 'Moving up the ladder or stuck at the bottom? Homeownership as a solution to poverty in South Africa'. *International Journal of Urban and Regional Research*, 35: 57–77.

Lodge, T. (2003) *Politics in South Africa: From Mandela to Mbeki*. Cape Town: New Africa Books.

McDonald, D. (2008) *World City Syndrome: Neoliberalism and inequality in Cape Town*. London: Routledge.

Mongwe, R. (2011) *Race, Class and Housing in Post-Apartheid Cape Town: What are the challenges of housing development initiatives in post-apartheid South Africa?* Pretoria: Human Sciences Research Council.

NDRC (National Resources Defense Council) (2013) *From Copenhagen Accord to Climate Action: Tracking national action to curb global warming*. Available online at http://www.nrdc.org/international/copenhagenaccords/ (accessed 1 November 2013).

News24 (2011) *R58bn needed to fix badly built houses*. Available online at http://www.news24.com/SouthAfrica/News/R58bn-needed-to-fix-badly-built-houses-20110210 (accessed 1 November 2013).

Parnell, S., Beall, J. and Crankshaw, O. (2005) 'A matter of timing: African urbanization and access to housing in Johannesburg'. In Brycson, D. and Potts, D. (eds) *African Urban Economies: Viability, vitality or vitiation?* London: Palgrave Macmillan, pp. 229–51.

Pharoah, R. (2011) 'Fire risk in informal settlements in Cape Town, South Africa'. In Pelling, M. and Wisner, B. (eds) *Disaster Risk Reduction: Cases from urban Africa*. London: Earthscan, pp. 105–24.

Satterthwaite, D., Huq, S., Reid, H., Pelling, M. and Romero Lankao, P. (2007) *Adapting to Climate Change in Urban Areas: The possibilities and constraints in low-and middle-income nations*. London: IIED.

Silver, J., Phillips, C. and Rowswell, P. (2011) *Mamre Ceiling Evaluation: Energy retrofitting in low income communities*. Cape Town: ICLEI Africa Secretariat and City of Cape Town Sustainable Livelihoods.

Sustainable Energy Africa (2010) *Urban SEED Update*, 2(6).

Toulmin, C. (2009) *Climate Change in Africa*. London: Zed Books.

UN-Habitat (2008) *State of the World's Cities 2008/2009: Harmonious cities*. London: Earthscan, p. 72.

Wlokas, H. and Corbera, E. (2009) 'Development benefits in the Kuyasa low-income housing CDM project in South Africa'. Paper presented at the *Earth System Governance: People, places and the planet* Conference, 2–4 December, Amsterdam.

World Bank (2010) *World Development Report 2010*. Washington, DC: World Bank.

7

REALIZING THE SOLAR CITY IN BERLIN, GERMANY

Introduction

In this chapter we turn our attention to Berlin and the creation of a Solar Atlas designed to stimulate the installation of solar photovoltaic (PV) and solar thermal (hot water) systems in the city. The project was initiated by the Berlin Senate, funded through the European Fund for Regional Development, project-managed by Berlin Partner (the city's PR arm) and delivered by virtualcitySYSTEMS, a company specializing in 3D city models. Conceived as a means through which to encourage investors and residents to imagine the solar potential of the city, the Solar Atlas required the alignment of actors at multiple institutional scales, and developed due to Berlin's growth and development ambitions, paired with the growing imperative at national and municipal levels to address climate change and energy security.

After briefly setting out the context of climate change policy in Germany and the urban context in Berlin, we examine the making, maintenance and living of the Solar Atlas. The Solar Atlas was successfully assembled, rendering the problem of climate change technical by showing the solar potential of the city, and enrolling available technologies, such as 3D city models and key interests. However, it has not been successfully maintained, and so far remains distant from the everyday practices of Berliners. The experiment therefore raises questions about how to govern interventions which are predominantly technical and, as a result, do not engage with the processes of subjectification, which has so far in this book proven so important to making experiments sustainable over time. Despite a lack of real maintenance or living, the Solar Atlas proves to be a productive window into broader renewable energy and urban development debates in Berlin, including the development of performance contracting for energy efficiency, where Berlin does emerge as a major innovator and leader. We argue that much of the innovation in Berlin has emerged as a result of the city's famously perilous finances, but that Berlin's poverty (both in financial and institutional terms) remains its greatest challenge as it seeks to reposition itself as a green energy leader.

Cities, climate change and renewable energy policy in Germany

Germany's commitment to climate change has materialized in numerous policy and technical interventions that have situated some of its cities at the forefront of climate governance. Freiburg, for example, calls itself a Green City and is well known for its very visible interventions in solar energy, as well as its capacity to mobilize multiple actors for climate action. At the regional scale, existing economic alliances have been mobilized to address climate change, as is the case in the Ruhr region. However, Berlin lags behind other German cities, in part because of its economic context. The Berlin Solar Atlas thus represents an effort to place the city firmly in the ranks of progressive cities in the German context. In this section we explain how climate change has become a national priority in Germany, and then examine how that has been translated into the local context of Berlin.

Climate change as a national policy priority

The importance of having a credible plan to develop renewable energy in Berlin can be explained by the wider context in which Germany has promoted itself as an international agenda-setter on climate change action, ever since Chancellor Kohl (of West Germany) stated in 1987 that the climate issue was the most important environmental problem facing the world (Lauber and Mez 2004). Since reunification in 1990, Germany has backed up its rhetoric with action at the Federal level. In 1990, the Government pledged to reduce CO_2 emissions for West Germany by 25 per cent of 1987 levels by 2005, revised in 1995 to 25 per cent of 1990 levels as a result of the changed situation with the newly re-unified Germany (Weidner and Mez 2008). Germany has also taken on a significant share of the European Union's common GHG reduction target. In Kyoto, the European Union (EU) agreed to an 8 per cent reduction of its 1990 GHG emissions, and Germany agreed to a 21 per cent reduction of its own GHG emissions to contribute towards this aim (Weidner and Mez 2008). Climate change action in Germany is structured around three basic actions: first, a shift to renewable energy sources; second, improvements in energy efficiency (particularly in housing and transport, and through expansion of Combined Heat and Power – CHP); and third, emissions controls through the EU's Emission Trading System (Weidner and Mez 2008). These policies have primarily been driven by the Federal government in a top-down fashion, but the generally more conservative States have been on-board, understanding that supportive national standards support their own renewable sectors (Weidner and Mez 2008).

Without colonies or ties to former colonies through which to secure a secure oil supply, the oil shocks of 1973 and 1979 led to serious concerns about energy security in Germany (see Lauber and Mez 2004; Weidner and Mez 2008). Whereas neighbouring France opted to resolve its energy security dilemma by investing heavily in nuclear power, this became an impossible option to sell to the German public following the Chernobyl meltdown of April 1986 in the nearby USSR (now Ukraine). In this context, renewable energy in Germany has emerged both as an imperative to create energy security, a pillar of climate change policy, and an avenue for the expansion of German industrial prowess (Lauber and Mez 2004; Weidner and Mez 2008). At the same time, the prominence of renewable energy in German climate change policy reflects the dominant view that Germany's vulnerabilities to climate change are economic.

The development of renewable energy sources has been encouraged through the use of a strong feed-in tariff system facilitated by the *Erneuerbare Energien Gesetz 2000* (EEG, 'Renewable Energy Act'). Amended in 2004, the EEG raised the feed-in rates for PV electricity to levels which made installations commercially viable without any additional support (Lauber and Mez 2004). This left a legacy of significant investment in solar generation, but has also meant that the government has had to gradually reduce the feed-in tariff being offered (Interview 1, Consultant, Berlin, January 2012). In addition to feed-in tariffs, Germany has pioneered the use of incentive programmes to promote the installation of solar generation. The 1989 *1,000 Solar Roofs* programme provided successful applicants with 70 per cent of funding costs of PV installations (50 per cent from the federal government, 20 per cent from the state) between 1991 and 1995, resulting in 5MW of total installations (Lauber and Mez 2004). In 1999, a *100,000 Roofs Programme* was adopted to promote growth in the number of solar installations. It provided reduced-rate loans for PV installations through the government-owned *Kreditanstalt für Wiederaufbau* (KfW) bank which, together with the feed-in tariffs, resulted in 350MW of PV installations (Weidner and Mez 2008). Along with other programmes for CHP and biofuels, this programme saw Germany achieve its 2010 target of deriving 4.2 per cent of primary energy supply from renewable sources by 2000. In line with the EU's wider ambitions, Germany's Energy Policy of 2010 sets targets for energy efficiency of a 20 per cent reduction in primary energy consumption by 2020 and a 50 per cent reduction by 2050, both based on a 2008 baseline. At the same time, GHG emissions must be reduced by 40 per cent by 2020 and between 80–95 per cent by 2050 based on a 1990 baseline. By 2020, 18 per cent of final energy consumption must be derived from renewables, increasing to 60 per cent by 2050. In the electricity sector, renewables must supply 30 per cent of power consumption by 2020, and for heating the share of renewables must be increased to 14 per cent by 2020 (Interview 2, Municipal Government, Berlin, February 2012). Renewable energy is therefore at the heart of current energy and climate change policy in Germany, in political, economic and cultural terms.

Climate change, energy and the urban context in Berlin

The importance of addressing climate change and of promoting renewable energy has also long been recognized in Germany's cities (Bulkeley and Kern 2006).

However, despite its political and cultural prominence, Berlin has not been a leader on these fronts. With a population of approximately 3.5 million, Berlin is the capital city of Germany and also one of the 16 states which together constitute Germany's federal system of government. Berlin's recent history is tightly bound up with its role in the post-World War II power-plays which developed into the Cold War. The Berlin Wall, progressively constructed from 1952 onwards, split the city in half, isolating the Allied-controlled West Berlin from the rest of the city, which was the capital of the newly-declared German Democratic Republic (GDR). One of the most significant infrastructural legacies of the division was the separation of Berlin's electricity, gas, water and wastewater networks (Moss 2009). Prior to World War II, Berlin had been particularly dependent on coal-fired electricity, and much of Berlin's electricity generation capacity was still serviceable or was quickly made serviceable at the conclusion of the war. However, the Soviets dismantled much of this infrastructure on their arrival in Berlin, including the largest and most modern plant, *Kraftwerk West* (Moss 2009).

Following the division of the city, network managers in both East and West Berlin continued to operate according to the supply-oriented logic which had been dominant in the city since the late nineteenth century. Infrastructure was used by both sides:

> as a means of protecting their own territorial integrity. The notion of infrastructure as a force for spatial cohesion and urban development was retained, but the spatial terms of reference were the newly truncated territorial units, not the city as a whole.
>
> *(Moss 2009: 935)*

The majority of the city's electricity generating capacity was located in East Berlin, along with important coal reserves. West Berlin, on the other hand, had some 70 per cent of the city's gas generation and storage capacity. Negotiating cross-border transfers of electricity and gas was a constant bone of contention between the administrations of West Berlin and the GDR, though by 1950 West Berlin's physical network of gas was almost completely independent of the East, and by 1995 West Berlin was self-sufficient in electricity generation.

However, when the Wall fell in 1989, reunification of these infrastructure networks was rapid, with networks reconnected within weeks and the utilities re-combined in just a few years (Moss 2009). The legacy of the city's division was oversupply due to duplication. Simply by recombining the networks, Berlin reduced its CO_2 emissions by over 25 per cent between 1990 and 2005, despite only 0.13 per cent of the energy mix for heat supply being derived from renewables (Interview 2, Municipal Government, Berlin, February 2012). However, a second legacy was that the city's significant CHP network, consisting of nine plants and supplying 50 per cent of Berlin's power, is dependent on fossil fuels, notably coal from the adjacent state of Brandenburg. Berlin therefore saw a clear need to develop renewable energy sources, and lacking the land required for wind power,

FIGURE 7.1 Urban redevelopment in an area where the Berlin Wall formerly ran, with a power station in the background. Photo © G.A.S. Edwards.

solar energy provided one of its few realistic options for reducing its dependence on imported fossil fuel-based energy. At the same time, large amounts of land had been taken up by the Wall and associated infrastructure dividing the city, and infrastructure had fallen into decay, so reunification led to a massive programme of infrastructure renewal in the city more broadly, creating opportunities for inserting new energy technologies in the city (Figure 7.1). In social terms, the city was transformed from a pre-war industrial city into a post-war cultural one, with mixed implications for socio-environmental relations. As one respondent reflected: 'From the point of emissions it's quite OK [laughs] from the point of economic welfare perhaps not' (Interview 3, Academic, Berlin, January 2012).

Berlin did not form part of the first wave of German municipalities who pioneered urban responses to climate change, preoccupied as it was with the politics and economics of reunification, and benefiting from the GHG emissions reductions that this process produced. But climate change remained high on the national policy agenda, and energy security remained an ongoing concern, with the effect that renewable energy continued to grow in importance. As Berlin sought to reinvent itself, this focus on renewable energy (and particularly solar) provided new opportunities to intervene in the energy infrastructure of the city. We explore this in more detail below.

Virtual experimentation? The Berlin Solar Atlas

The Berlin Solar Atlas is an online map launched in 2007 to allow residents and investors to investigate the potential for solar energy across the city and determine the best rooftop locations in which to develop solar PV and thermal installations. Its proponents hoped to promote the development of solar energy in the city, which has considerable potential but below average penetration and, as the previous sections highlighted, has historically depended primarily on fossil fuel-based energy sources. The Solar Atlas was nested within a much larger project to create a 2D and 3D model of the city as a one-stop-shop for investors who were considering locating their business in Berlin. It emerged, then, at the intersection of the dynamics of growth and development in the city, and the wider policy context in which issues of climate change and renewable energy were growing in importance. As we argue below, this context facilitated the establishment of the Solar Atlas as an experiment in changing how the city is viewed and how renewable energy is problematized, but maintaining the Solar Atlas and embedding it in the everyday practices of actors involved in the solar sector – including manufacturers, installers, building managers and residents – have proven much more challenging.

Energy contracting and the making of the Solar Atlas

The origins of the Solar Atlas arguably lay in the mid-1990s, when severe pressure on Berlin's city budgets coincided with broader European shifts towards liberalization of energy markets. In Berlin, this led to the privatization of the city's energy utilities. The Berlin Senate sold its majority shareholding in the electricity utility BEWAG in 1997 and the gas utility GASAG in 1998, both to transnational consortiums (Monstadt 2007). The Senate compelled the new owners to adopt public interest obligations through contractual agreements, but the privatizations nevertheless had the effect of reducing the power of the Senate to control the corporate policy or investments of the utilities: in 2001, the new private owners of BEWAG sold their stake to the Swedish state-owned energy company Vattenfall, which had already taken over the two largest East German energy companies and Hamburg's utility. BEWAG gave up its autonomy in the process, being incorporated into an entity called 'Vattenfall Europe' (Monstadt 2007), and the Senate largely lost its ability to impose renewable energy targets through the utility.

But if the financial difficulties experienced by Berlin in the mid-1990s led to the Senate losing a degree of control over the utilities in the process of privatization, they also underpinned the most significant climate change policy innovation in the city: Energy Saving Partnerships (ESPs). ESPs seek to deliver both emissions reductions and cost savings through energy saving retrofits of publicly-owned buildings. ESPs involve grouping government-owned buildings together, and then allowing the private sector to tender to install technical improvements in the buildings. The financial savings associated with the energy efficiency improvements are then shared by the contractor and the Berlin Senate. Contractors must guarantee energy savings of over 15 per cent over 12–15 years, a model known as performance contracting. As of 2012, 24 ESPs covering 1,500 buildings had delivered Berlin's government with savings of €3.2 million per year (Interview 2, Municipal Government, Berlin, February 2012).

As with the privatization of the utilities, the emergence of ESPs (colloquially referred to as 'contracting') and the technical and institutional expertise required to administer them has created a new landscape of power with respect to the city's energy and climate change politics. One particular actor, the Berlin Energy Agency (BEA), has acquired particularly noticeable financial and intellectual power. Founded in 1992 (and with capital stock of just over €2.5 million from the privatization of the electricity utilities), the BEA is an energy services consultancy focused on energy efficiency and renewable energy, providing both information and technical expertise (Interview 1, Consultant, Berlin, January 2012) (see also Monstadt 2007). It has four shareholders, namely Vattenfall (the electricity utility), GASAG (the gas utility), the Berlin Senate and the KfW Bank. Apart from its capital stock, BEA does not receive public funding, but its close connections to both the government (one respondent explained that many of the BEA's staff were in fact drawn from the Senate) and the major privately-owned utilities clearly provide a significant competitive advantage, and it has grown rapidly, expanding from 10–15 staff in 2002 to over 60 in 2012 (Interview 1, Consultant, Berlin, January 2012; Interview 4, Journalist, Berlin, February 2012). Developing ESPs with and on behalf of the government is core business for the BEA, and a representative explained that the BEA 'have good standing I think we can say worldwide for the energy performance contracting which is something we have really developed here in Berlin together with the Senate and so-called energy saving partnerships' (Interview 1, Consultant, Berlin, January 2012). Critically, the BEA sees itself as a thought leader, a key task of which is to 'tell the state government what they should be doing' (Interview 1, Consultant, Berlin, January 2012). The BEA business model is to target the economic levers encouraging energy efficiency and the uptake of renewable energy in the city. BEA is a particularly privileged manifestation of what Monstadt calls 'ecopreneurs'. According to Monstadt:

> Due to the innovational lethargy of the regional utilities and the inefficiency of the traditional regulatory system, environmental policy in Berlin has increasingly focused on the promotion of innovative SMEs specializing in the production and use of environmental technologies, in the ecological supply of electricity and heating, and in environmental services. These 'ecopreneurs' ... have become indispensable promoters of industrial transformation and an ambitious climate policy.
>
> *(Monstadt 2007: 333)*

Thus, top-down efforts to forge alignment for energy innovation has turned attention to 'ecopreneurs' as agents of change. This is not just a coincidence, but rather an effective way through which the will to improve is displaced towards actors who take an active role in producing and enacting certain forms of rationality. Moreover, the ESP scheme has been supplemented by the Senate seeking voluntary agreements with companies to save energy and/or reduce their CO_2 emissions at no cost to the government (Monstadt 2007). Thirteen of these voluntary agreements have been signed with companies including the gas utility GASAG, waste and water management companies, a real estate management company and many of Berlin's municipal housing companies, including the *Verband Berlin-Brandenburgischer Wohnungsunternehmen e. V.* (BBU), a municipal housing association which is responsible for 40 per cent of Berlin's flats. However, the most significant voluntary agreement signed was with the electricity utility Vattenfall, which has committed to converting three of its CHP plants from lignite-fired to either gas- or biomass-fired. These agreements can be understood as strategic attempts to align the interests of the state and private enterprise in the name of climate change and cost savings. They seek to render the problem of energy inefficiency technical by focusing on the machinery which can improve energy efficiency, but in setting up these technical projects they also reconfigure the nature of climate governance. In particular, they shift agency and power from the hands of public institutions to private companies.

The culture of performance contracting through ESPs supplemented by voluntary agreements is vital for understanding the Solar Atlas. In 2002, Berlin's Senate established a performance contracting scheme called the *Solardachbörse* ('Solar Roof Exchange') in an attempt to stimulate development of solar energy in the city. According to a representative from the *Senatsverwaltung für Stadtentwicklung und Umwelt* ('Senate Department for Urban Development and the Environment'), the intention of the *Solardachbörse* had been to start with government-owned buildings, but subsequently expand the programme to include privately-owned roofs (Interview 2, Municipal Government, Berlin, February 2012). However, when applied to solar energy, the performance contracting model encountered challenges which had not arisen in the context of energy efficiency upgrades and CHP installations. One of the issues, according to an industry source (Interview 5, Solar Entrepreneur, Berlin, February 2012), was reluctance even amongst government building-owners to release their roofs to contractors on the terms and for the durations investors considered necessary. As a result of the contractual impasse (further compounded by questions about liability should something go wrong with an installation), the *Solardachbörse* foundered. It demonstrated the clear interest in developing solar energy in the city, but the detailed calculations required to implement the projects drove apart actors whose interests were in broader terms aligned. It was clear that some other mechanism would be required to stimulate solar installations in Berlin.

That mechanism was the Solar Atlas, which sought to make solar installations more amenable to investment by visualising the potential for such installations across the city. The opportunity for developing an atlas emerged from a strategy being undertaken to encourage the economic redevelopment of the city. During 2007, a city model was developed in 2D and 3D to provide an interactive virtual tour of the city, which includes 'economic data, infrastructure, political structure … labour markets and … real estate' (Interview 6, Project Manager, Berlin, December 2011), for potential investors, in line with Berlin's desire to promote the economic revitalization of the numerous localities which had decayed through the years of the city's division. Though it can be accessed online, Berlin Partner promotes it particularly through the

Business Location Center, a board-room in central Berlin which potential investors are encouraged to visit as their first point of call in Berlin.

The 3D city model was built based on Light Detection and Ranging (LIDAR) data, collected from a purpose-flown aerial survey in 2007 with a resolution of 50cm pixels. The idea of integrating the solar potential of different rooftops in the city with this data emerged after Berlin Partner 'started to think well what else can we do with the LIDAR data and so at this point of time several companies started to offer services for analysing the LIDAR data for solar and for suitability for further installations' (Interview 6, Project Manager, Berlin, December 2011). virtualcitySYSTEMS, a company specializing in the development of 3D city models and the delivery agent for the city model, first developed a pilot consisting of two sample areas of 20km^2 each, 'one was in the city centre and one was more in the suburbs with different kind of living structures' (Interview 6, Project Manager, Berlin, December 2011). This pilot suggested that 'it's not only useful to show roofs which are suitable for solar installation but also to give maybe positive examples of already existing installations and it's possible to visualize the suitability for photovoltaic installation and for permanent installations' (Interview 6, Project Manager, Berlin, December 2011). Funding for this additional layer of the city model was provided in part by the European Fund for Regional Development, and came with the condition that it 'had to be something that is open to the public so that everybody can use it' (Interview 6, Project Manager, Berlin, December 2011).

From this conjunction of the city model and the potential for using the LIDAR data to different ends, the Berlin Solar Atlas was created to highlight the potential for both Solar Photovoltaic (PV) and Solar Thermal installations on building rooftops throughout Berlin in the 2D model and within Berlin's S-Bahn ring in the 3D model. As described on its website:

> The Berlin Solar Atlas shows the solar potential of each building in the city in clear, sharp images in both 2D and 3D. Property owners and investors can use the Atlas to determine whether a building's roof is suitable for a solar installation and whether the investment will pay off. The Atlas provides key information at a glance on such matters as the potential power output, reductions in CO2 emissions, and investment costs.
>
> *(Business Location Center n.d.)*

The Berlin Solar Atlas also shows the location of existing solar installations in the city, a selection of solar-related organizations, companies, research facilities and higher education institutions, and government-owned buildings. According to one of it its designers, the purpose of the Solar Atlas is to encourage those interested in installing solar to pursue it further by providing easy-to-access information about possible electricity yield, CO_2 savings, installable panel sizes and estimated cost:

> it's for people who are who have a basic interest in having solar panel on their roof and it's just there for giving a first impression of what might be

> possible so it's just to give them a rough idea if it's worth maybe getting a specialist to inspect the roof properly.
>
> *(Interview 6, Project Manager, Berlin, December 2011)*

Developing the Berlin Solar Atlas was innovative from a technical perspective, because it involved developing techniques for analysing roof slopes and solar potential, combining this with shadow analysis, overlaying these calculations with estimates of solar radiation, and then categorizing, mapping and visualizing this data for presentation on both a conventional, 2D map layer and on the 3D city model (Figure 7.2). The representative from virtualcitySYSTEMS explained that the raw data was analysed by a contractor 'and we had to visualize the data for both the public and also the visitors here in Business Location Center' (Interview 6, Project Manager, Berlin, December 2011). This involved a complex process of calculation. They classified the buildings into four classes, according to their potential for solar PV installations. The 2D data with all the solar-related attributes supplied by the sub-contractor then had to be integrated into both the 2D map and the 3D city model: a significant technical challenge because it required linking three large and complicated databases – one containing the 3D buildings, another containing the 2D city map, and the third containing the Solar Atlas data. Ensuring the Solar Atlas layer was correctly rendered on the rooftops of the 3D buildings involved not only classifying buildings according to solar potential but also geo-referencing the layer, such that the roofs appeared at the right altitude and angle. So from a technical perspective the Solar Atlas was considered a success, and 'the politicians liked it and because Berlin wants to represent itself as a new, as a green city that does a lot for solar energy' (Interview 6, Project Manager,

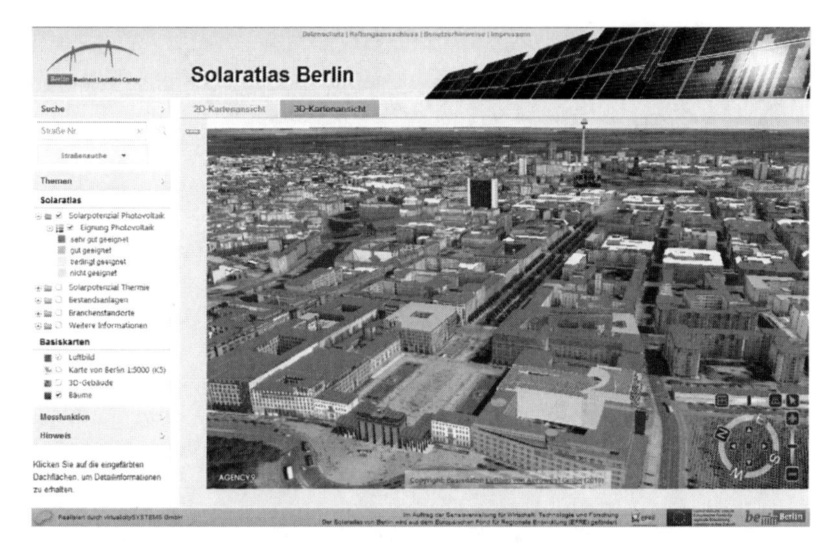

FIGURE 7.2 An image from the Solar Atlas. Image © Berlin Partner/virtualcitySYSTEMS GmbH, used with permission.

Berlin, December 2011). The focus here was in aligning economic and technical criteria through assembling the instruments that would make the solar potential of Berlin visible, and thus amenable to investment. The social aspects, from how it would be publicly adopted and the institutional and policy arrangements that would make it possible, were largely overlooked.

Aligning the energy system with the emerging economy of actors and institutions invested in the redevelopment of the city, the Solar Atlas was framed as a means through which to achieve new forms of contract between private investors (both commercial and residential) and the infrastructural development of Berlin. In this manner, rendering Berlin governable as a space for the development of solar energy was predominantly conducted through processes of calculation, of gathering, translating and conjoining information about the solar potential of the city in technical terms. As the Berlin Solar Atlas took shape, the uptake of solar energy in the city was problematized as a matter of providing sufficient information so that developers and residents could make reasoned decisions about the potential and payback of their potential investments. The challenges which had stalled the *Solardchbörse* were left to one side, as were the ways in which investment in solar energy might gather more emotive force.

Incomplete maintenance

As we argue in Chapter 2, experimentation does not rest on the assembly or making of particular forms of intervention, but requires processes of maintenance, which serve to sustain the experiment and embed it within the wider metabolic flows of the city. Processes of upkeep, we suggest, are the means through which the experimental qualities of any intervention are both made to endure and at the same time become normalized within the particular urban context in which they emerge. The Solar Atlas, by its very nature, stands as a snapshot of solar potential at a particular time. Upkeep in the form of updating the underlying data and map data is therefore essential for the ongoing utility of the Atlas, particularly in Berlin, where the rate of building renovation has been notably high since the fall of the Wall. To give some indication, one respondent (Interview 6, Project Manager, Berlin, December 2011) estimated that in the two years between 2007 and 2009, approximately 10,000 buildings in Berlin had changed in some form or other. Without a plan and programme for updating the data presented in it, the field of intervention created by the Solar Atlas becomes both smaller and more unstable with each passing year, as the processes of reconfiguration and circulation in the city mean its fabric – and consequent solar potential – is constantly being reshaped. However, this upkeep has not occurred: respondents indicated that there are no plans to update the base LIDAR data on which the 3D city model and Solar Atlas are based, despite the fact that at the time of fieldwork in 2012 it was five years old. This also raises questions about how the problem of fostering investment in energy innovation was rendered technical, in relation to whether it was realistic to think that the Solar Atlas would be perpetuated in time or it was to be understood as a one-off intervention. The potential for urban reconfiguration of a one-off intervention, without given mechanisms for the enrolment of wider publics, is very small.

If upkeep has been entirely absent in a technical sense, attempts have been made to promote the Solar Atlas as a novel means through which to engage residents and developers with the idea of solar investment. A respondent from the BEA reflected that: 'it's just more or less a communicational approach to get people interested in that [whether it's possible to install PV] and to see OK can I use my own roof' (Interview 1, Consultant, Berlin, January 2012). Initially, according to one solar entrepreneur (Interview 7, Solar Entrepreneur, Berlin, February 2012), the Solar Atlas was promoted at trade fairs to industry professionals, rather than to building owners. This proved to be the wrong audience, because such solar entrepreneurs already knew about the potential of solar installations and saw little utility in the way the Solar Atlas mapped solar potential because the level of data included was not sufficient for the exact calculations required at the point of installation: 'you still need to do your own calculations, you still need to see or look out for so many different things that are not in there' (Interview 5, Solar Entrepreneur, Berlin, February 2012). In short, the promotion of the Solar Atlas at trade fairs was preaching to the converted, missing out on its considerable potential for raising awareness about solar potential amongst the broader population and building owners in particular. In contrast, a similar project in Osnabrück had been supported by a large-scale communication exercise where:

> they wrote a letter to every house owner where the rooftop was suitable for solar installation and offered them free information with a specialist and so in the end I don't know maybe ten per cent of the people finally installed an installation on their rooftop and so the city quickly reduced the city emissions.
> *(Interview 6, Project Manager, Berlin, December 2011)*

From the outset the Solar Atlas attracted a mixed response. Amongst those who did hear about it, some people thought it very useful, whilst others complained that their buildings were missing. Furthermore, there was no strategy for translating the interest in solar it generated into projects, such that the experiment lacked upkeep in terms of the continual circulation of its intention and potential. An environmental activist saw the creation of the Solar Atlas as a first step, 'but then to really to get active it would be very much better to have someone to speak to ... you need some campaign, you need information materials, you need to be connected to people to install this' (Interview 4, Journalist, Berlin, February 2012). Such strategies are particularly important in Berlin which, unlike Osnabrück, is dominated by multi-family housing, which makes connecting with building owners and motivating them to install solar more challenging, since it is more difficult to establish clear links between the costs of installation and the benefits they will derive.

The failure of the Solar Atlas to provide an adequate mechanism to translate information into action has been acknowledged by the developers of the Solar Atlas, who have subsequently engaged RDS Technologies (RDS) as an exclusive partner to promote the Solar Atlas. RDS acts as an intermediary between building owners and installers, and built up a database of 1,700 projects in the PV market, including investors and project development companies. In 2009 RDS started the *Dachkampagne* ('Roof campaign'), creating 'a transparent market to facilitate the installation' (Interview 7, Solar Entrepreneur, Berlin, February 2012) by connecting building owners directly and efficiently with pre-qualified PV installers. An interviewee from RDS explained that they can offer proposals from five installers within as little as three days, but that, despite this, there was 'not so much business in Berlin', where she thought there was more talk about solar than action. The fact that RDS was engaged to promote the Solar Atlas raises the possibility that the Atlas has served as a means to develop the field of intervention over which other – and very different – experiments (such as the *Dachkampagne*) have been deployed. Yet the RDS Technologies representative stated in relation to the Solar Atlas that 'I don't need them so much as they need me', suggesting that as experiments get consolidated, those actors whose experiments are (more) successful lay a stronger claim to the delivery of further strategies and calculations. Another solar entrepreneur, for example, stated that:

> the actual promotion of those roofs [in the Solar Atlas] didn't work and then RDS came in and said well we'll do the promoting for you. And so they basically promote what's already there for free on the market. They promote it and they got the more or less exclusive rights. So they just sit in the middle and make money.
>
> *(Interview 5, Solar Entrepreneur, Berlin, February 2012)*

This also speaks to the complex process of forging alliance during the making of an experiment and its subsequent consequences. Awarding the contract *post hoc* to one partner in this case excluded other possible partners. The challenge of inserting solar power systems into the flows of resource, finance and energy in the city have been considerable, such that the effect of the Solar Atlas in achieving any form of metabolic adjustment has been limited. To achieve a more fundamental realignment of such flows, we suggest the Solar Atlas would have needed to engage much more completely with the urban context in Berlin, and particularly with two intertwined structural constraints: the fact that it remains a predominantly poor city dominated by renters, and the political climate of the city. On the first point, the former Mayor of Berlin, Klaus Wowereit, famously observed in a TV interview in 2004 that '*Berlin ist arm, aber sexy*' ('Berlin is poor, but sexy'). Ever since reunification, with the deindustrialization it entailed for the city and the withdrawal of external funding which had supported the city during the city's division, public finances have been perpetually strained. But personal as well as public finances are typically strained in Berlin, where most people in Berlin rent their homes, giving rise to the well-documented problem of split incentives for investments in energy efficiency. In Berlin, this is exacerbated by the large number of old buildings with roofs that may not be suitable for solar installations or may need considerable remedial or upgrading work to make them suitable – a challenge it shares with Philadelphia (Chapter 5). As one respondent put it:

> in Berlin 80 per cent of the people are tenant[s] and a number of reason why we have not so many solar system on the roofs then you have a lot of old houses, old roofs which are not suitable for that.
>
> *(Interview 8, District Government, Berlin, January 2012)*

Where the city itself is not maintained, making space for solar energy becomes a fraught task. In fact this challenge runs deeper than merely the insertion of solar energy systems; it is also a problem for energy efficiency projects more generally, as one respondent explained:

> the one who is living in a flat is not the one who has to invest in the energetic refurbishment and re-optimisation of the building ... there is no real pressure on building owners to really do something [about energy efficiency] ... there is no legal requirements apart from the federal legislation.
>
> *(Interview 1, Consultant, Berlin, January 2012)*

There have been a number of attempts to introduce a law mandating certain levels of refurbishments, promoted by organizations such as BUND-Berlin and the Coalition of Tenants, but these have been continually postponed (most recently in 2010), primarily because German law allows 11 per cent of modernization costs to be levied onto the tenants directly. This 11 per cent levy does not provide an adequate incentive for landlords to make the investments, but is enough to be very problematic in social terms. As one respondent put it 'Berlin is a poor city or there are very many people who are poor so there is this problem with rising rents and some people think there is a conflict between social and ecological goals' (Interview 9a, Activist, Berlin, February 2012).

Clearly strong leadership is of critical importance in this context, but this has been notably lacking in Berlin with respect to climate change and renewable energy, which have remained low on the government's priority list. This political climate has formed a second key structural barrier to the ability for climate change experiments to gain traction. A retired researcher who had been deeply involved in *Klimaschutz* ('Climate protection') initiatives in Berlin explained that of the 100+ pages of the contract between the two political parties who together form the coalition government only:

> on three or four pages there is a little bit written about the climate protection policy but it's just of that kind to say we love peace and nobody knows what will be the real result of that, wait and see.
>
> *(Interview 3, Academic, Berlin, January 2012)*

The Senate administration itself has been subject to ongoing restructuring as the mirrored bureaucracies of the GDR and West Berlin have been dismantled. As part of this, jurisdiction for climate change and renewable energy has until recently been distributed across various departments, hindering effective policy-making and implementation. For instance, under the previous Berlin administration, energy policy fell into the Department of Economy's jurisdiction, 'but in reality they had nobody there who cares about energy policy' (Interview 3, Academic, Berlin, January 2012). More recently, a climate change group has been established within the *Senatsverwaltung Stadtentwicklung und Umwelt* ('Senate Department for Urban Development and Environment'), but it consists of just four people. Together, this institutional instability and lack of adequate staffing has meant that much of the work of policy development and implementation has been outsourced to contractors such as the BEA. In short, poverty and limited political interest have proved to be significant barriers to maintaining the Solar Atlas, and this lack of maintenance has in turn meant that the experiment has not achieved purchase in the city, and thus, it has not really been lived.

The limitations of bringing experiments to life

The Solar Atlas highlights the broader challenge for stimulating development of renewable energy in Berlin. Despite its aspirations of climate change and renewable energy leadership, Berlin has not been able to successfully translate rhetoric into results. There is little evidence that projects on the ground have developed out of either the *Solardachbörse* or the Solar Atlas, and in 2011 the proposed *Energy Saving Act* foundered once more on the floor of the debating chamber, despite eight years of effort from its proponents. The *Solardachbörse* and Solar Atlas were neither maintained nor really lived. Instead, they were appropriated by a few and, over time, their experimental character disappeared and lost legitimacy. This shows that just assembling an experiment is not enough. For instance, one respondent explained that the *Solardachbörse* ran into 'many problems with bureaucracy ... in many districts. Some districts were quite co-operative, I think Charlottenburg also, but the most make really problems. So there were many hurdles and obstacles to get permission [to include buildings in the *Solardachbörse*]' (Interview 4, Journalist, Berlin, February 2012). This was described by the representative from the BEA as a 'capacity gap' which prevented public building owners from demonstrating that their roofs are ready to hold PV installations, and so failed to qualify for inclusion in the *Solardachbörse*. A representative from one of the most progressive districts concurred, explaining that:

> we have some PV systems on school roofs. Five of them at the moment. But we had some problems with our roofs, we have some problems with static and we had some political problems because we had a problem with one school and then one party complained against the other party who wasn't charged for the school and so on.
>
> *(Interview 8, District Government, Berlin, January 2012)*

These political challenges raised the question of liability if something went wrong (mentioned above under *Energy contracting and the making of the Solar Atlas*), derailing the solar installations in that district. The respondent continued:

> … and so on, so we stopped that because our schools and our roofs are not the best and if you do something with your back at the wall and I don't take any responsibility then you can't do these things because and so what's happened if we have two meters snow and what's happened if there is the storm and then the solar panel will flew from the roof who is in charge, who paid and so on … there are always one thousand reason not to do something and only one reason to do it.
>
> *(Interview 8, District Government, Berlin, January 2012)*

An inherent barrier to the effectiveness of the *Solardachbörse* was the fact that, though public building owners did not have the financial capacity to install solar at present, they wanted to retain control of the roof for potential future self-built installations. One youth activist explained:

> I just wrote to my housing organization and it's a public organization it is owned by Berlin and they told me 'we don't want to build solar panels on our roof because we don't have the financial capacity but we also don't want to rent a roof to someone else' … because they want to have the possibility to build it themselves so they are just wasting the roofs [laughs].
>
> *(Interview 9a, Activist, Berlin, February 2012)*

Creating a field of intervention thus gives rise to strategies among those actors who accept the problematization and are enrolled in the calculations but recognize their lack of capacity to intervene without having their interests co-opted in those by other actors. The possibility of future intervention is here an obstacle for the present possibilities to make solar possible. Moreover, a solar entrepreneur explained that a number of factors meant public building owners were reluctant to sign performance contracts for solar installations:

> I think most people in solar would be brave but actually to give away the roof, the ownership of the roof for twenty years is something they don't want … and then you've got a lot of the times the council hall sees how much money could be made if they owned the system by themselves, they don't have the money to do it but they are like 'why aren't we getting a higher return?' [It's] because they don't bear any risks because they see how much money then they want, they get greedy, they want money so that's really hard and that's our experience especially here in Berlin.
>
> *(Interview 5, Solar Entrepreneur, Berlin, February 2012)*

As a respondent from Berlin's Ecologic Institute – who had been engaged to evaluate the ESP model – reflected, 'the goal was mentioned to expand that contracting concept to include solar energy on a regular basis but at least at the time of our evaluation this was not strongly implemented' (Interview 10, NGO, Berlin, February 2012). This suggests that the primary motivator for solar was financial rather than environmental or altruistic: 'I've never heard of someone who does it only for the climate, it's all about the money' (Interview 5, Solar Entrepreneur, Berlin, February 2012). The representative from the BEA likewise explained that, more broadly:

> you have to do something for climate protection but in the end it's always they ask 'OK what does this cost' and then you have to have the answer, you have to have the right financial models and you have to have good examples in hand which show it's reasonable and it's doable.
>
> *(Interview 1, Consultant, Berlin, January 2012)*

Thus, the strategic calculations of the different actors intervening necessarily need to include considerations about the particular moment and place of intervention, and this is translated by very particular geographies of innovation that relate to independent efforts to forge alignment around a particular interest.

In privately-owned buildings, a number of barriers were similar to those facing public buildings. Private owners were no more willing than public owners to commit to the long contracts required to make solar installations economically viable

> house owners want to be stay flexible and perhaps they want a big loft at some places or don't know if they would sell the house later on so they don't plan for 20 years or more to build something which will keep them, tie them to let their roof like it is.
>
> *(Interview 4, Journalist, Berlin, February 2012)*

The condition of roofs was also a factor, just as it was for the Coolest Block contest in Philadelphia (Chapter 5). A respondent from BUND-Berlin noted that 'If the roof is very old ... you first need to fix up the roof and then you can install it [solar] because it stays there for 20 years and it makes no sense to install first and then fix up the roof' (Interview 9b, Activist, Berlin, February 2012). But an additional barrier emerged, too: gaining agreement amongst the multiple owners in the building to rent their roof was far more difficult than for installations on the single-owner dwellings typical of smaller towns and villages (Interview 3, Academic, Berlin, January 2012), and this was not helped by the fact that the return-on-investment (ROI) period for solar in Germany was eight to nine years, while most investors in Germany sought a four to five year ROI.

Overall, there was a failure of relating the experiment with aspects of everyday life in Berlin. One determining factor has been the assemblage of the experiment as a technical issue, the rendering of the solar map in 2D and 3D, without a serious consideration of how this map related to any social, cultural or environmental dimension of living in Berlin. What is more puzzling, however, is that the experiment also fails to be lived by those technicians who could possibly engage with it in relation to their own everyday practices and, fundamentally, to their construction of narratives of energy intervention. Is not just that the Solar Atlas had little to say about how to live with solar energy in a particular home, but also it had little to say about how to live the development and implementation of energy technologies, and as a result has become a one off, momentary experiment that few actors in Berlin engage with.

Impacts and implications

It is clear that, though the Solar Atlas has been successfully assembled in Berlin, it has not been successfully maintained, and there is little evidence that it is being lived. Despite this, it proves extremely valuable as a window into contemporary climate change and renewable energy action in Berlin.

To start off, it has had a number of very positive effects. One of these has been the fact that it has helped to create a new culture of Berlin as a 'green city' which, given Berlin's history, is an important agenda to create momentum behind. However, it also shows how important carbon/energy calculation is in experimentation. The technical expertise required to assemble and create the Solar Atlas shows just how dependent experimentation is on new techniques, and even the development of a whole set of professional roles associated with this: not only were government officials and project managers involved, but also energy agencies, intermediaries who campaigned for solar, entrepreneurs running solar businesses, and even the installers doing the actual solar installations. These diverse actors are bought together in the process of assembling the experiment, but part of why the experiment failed to become lived in Berlin was the fact that they all have different forms of calculation at work, and these calculations were not made commensurate with one another in the Berlin case.

Second, the Solar Atlas exists in two dimensions (though it is rendered in three), at a particular junction in time. As it turns out, the smooth space of the Solar Atlas which purports to represent reality, but in fact represents a particular vision of a reality captured at a particular point in time, actually obscures the complexity of maintaining and living. It serves to iron out difference, or to calculate those differences in terms of solar potential and financial return, but the differences of the city in terms of ownership, building fabric and existing infrastructure connections and dependencies are lost.

Third, the reason the experiment fails to be lived can be understood in terms of sticky circulations (much like sticky keys on a keyboard): the Solar Atlas highlights the challenges that can arise if experiments are not rendered compelling to the right 'circles', be they political, financial, business, or even citizen. With nowhere to go to, the Solar Atlas ultimately failed to create a new form of economy, despite the attempts to rescue it by bringing RDS on board to promote it. Even in the context of economic subsidies and the other solar programmes, these circuits are all moving in different directions, and the Solar Atlas could not find traction in any one of them, and hence remains moribund and symbolic, failing to gain purchase on the solar resources of the city in any transformative sense.

Finally, the experiment's singular focus on the technical left little room for subjectification of the most appropriate sectors of society. There were some attempts to assemble its social, cultural components, but these fell foul of the very significant resource needs to provide upkeep for such a project, let alone metabolic adjustment. In a sense, implementation of the Solar Atlas perhaps required a much more cybernetic project of social control, one in which technologies are rolled out if only you have all the data of the city and you can update it in real time. This would also occur through a process of subjectification on a much larger scale. Subjectification, as it turns out, was one of the defining weaknesses of the Berlin Solar Atlas, which was notable for failing to engage with either the citizenry or the solar industry. Failing to provide a compelling account of what low carbon living means in the context of Berlin ultimately meant that, despite its transformative potential, the experiment was relegated to a dusty server room.

References

Bulkeley, H. and Kern, K. (2006) 'Local Government and the Governing of Climate Change in Germany and the UK'. *Urban Studies*, 43(12): 2237–59.

Business Location Center (n.d.) *Berlin Solar Atlas*. Available online at http://www.business-locationcenter.de/en/berlin-economic-atlas/the-project/project-examples/solar-atlas (accessed 6 January 2014).

Lauber, V. and Mez, L. (2004) 'Three Decades of Renewable Electricity Policies in Germany'. *Energy & Environment*, 15(4): 599–623.

Monstadt, J. (2007) 'Urban Governance and the Transition of Energy Systems: Institutional Change and Shifting Energy and Climate Policies in Berlin'. *International Journal of Urban and Regional Research*, 31(2): 326–43.

Moss, T. (2009) 'Divided City, Divided Infrastructures: Securing Energy and Water Services in Postwar Berlin'. *Journal of Urban History*, 35(7): 923–42.

Weidner, H. and Mez, L. (2008) 'German Climate Change Policy: A Success Story With Some Flaws'. *The Journal of Environment & Development*, 17(4): 356–78.

8

SOLAR HOT WATER AND HOUSING SYSTEMS IN SÃO PAULO, BRAZIL

with Andrés Luque-Ayala

Introduction

Solar hot water (SHW) experiments in social housing are creating a new model for urban infrastructure and energy provision in the State of São Paulo, Brazil. In a country where 80 per cent of dwellings heat water for showering and 73 per cent rely on electric showers for this purpose (PROCEL/Eletrobras 2007), SHW systems have the potential to provide a significant contribution towards a reduction in greenhouse gas (GHG) emissions, whilst also balancing and optimizing electricity demand patterns. This chapter is based on the experience of the Housing and Urban Development Company (CDHU, *Companhia de Desenvolvimento Habitacional e Urbano do Estado de São Paulo*) of the State of São Paulo in promoting the use of SHW in its dwellings. Although Brazil's electricity generation is largely based on hydropower (a renewable energy source), SHW is seen by both the State and the municipality as playing a relevant role in their carbon reduction strategies. However, narratives around the use of SHW are taken beyond sustainability discourses and into the space of social justice and poverty alleviation, as they become pivotal in furthering a discourse on 'housing with dignity'. The chapter demonstrates how experimentation, through the use of small material agencies of technological objects, serves to interweave climate change as a means through which to address other social concerns in the city.

Through the case of CDHU's experimentation with SHW, this chapter analyses the interaction between the social and material agencies involved in achieving low carbon in the city. It argues that making, maintaining and living low carbon requires acknowledging the role of the seemingly small material agencies that shape both policy responses and user practices. However, rather than focusing on the practices of consumption at play in low (or high) carbon living (Shove and Walker 2007), it looks at the practices of production of low carbon energy, primarily through the perspective of the socio-material agents driving change in how electricity is used in social housing. These agents form part of an assemblage, a fragile, contingent and emerging socio-spatial arrangement (Anderson and McFarlane 2011; Bennett 2010) pushing for a greater adoption of SHW systems with a multiplicity of overlapping motivations, from climate change to the optimization of Brazil's electricity grid and the fulfilment of social agendas via sustainability innovations. Along with the CDHU, these actors include the regional electricity utility companies, SHW dealers and manufacturers, construction companies, armies of architects and engineers, and users. At the same time, the analysis points to the critical role of material agencies (amongst others, water as a primary medium for the transfer and use of energy in housing, tower blocks as the building typologies imposing limitations on the experiment, and a group of seemingly small and simple technological objects such as electric showers and SHW systems) that together serve to reconfigure the ways in which energy services are provided and, in turn, what such services mean. Such materials are enrolled in the assemblage through specific calculative practices that ultimately determine not just how the experiment is assembled, but also how it comes to be maintained and lived over time.

Urban development in São Paulo

With a population of over 40 million, the State of São Paulo has historically been Brazil's main centre of economic activity. Its capital, the City of São Paulo, sits at the centre of the second largest metropolitan region in the Americas. Adjacent to it, two other metropolitan regions make up the large majority of urban São Paulo: Campinas and the Baixada Santista. Together, these three metropolitan regions

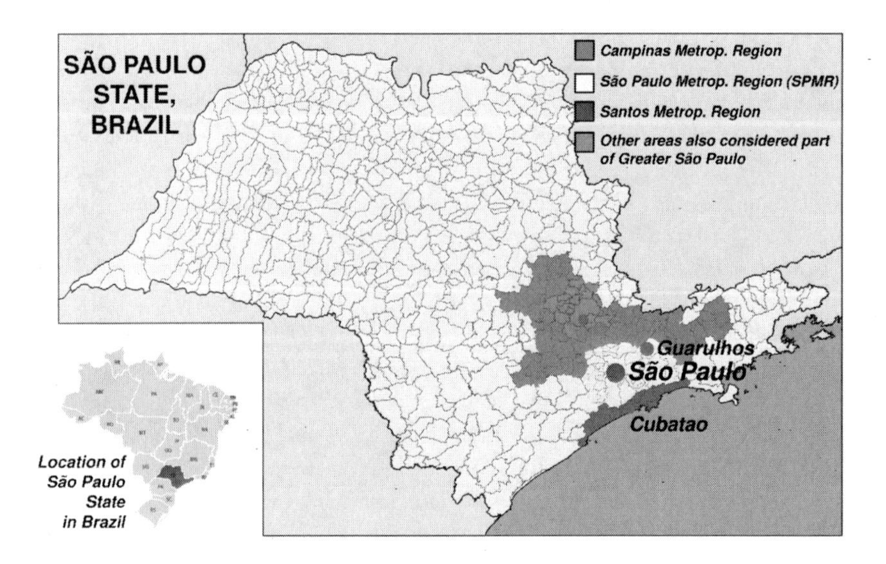

FIGURE 8.1 Map of São Paulo. Drawn by A. Luque-Ayala, used with permission.

account for an estimated 25 million people and over 67 cities, one of the largest urban agglomerations in the world (Figure 8.1). The largest of these cities, the City of São Paulo, has often been praised for the implementation of a pro-poor and pro-inclusion approach to urban development, where progressive policies have been effective at providing the urban poor with greater access to educational opportunities, housing, bank credit, transport and recreational opportunities (UN-Habitat 2010).

Providing social housing

Despite the significant efforts invested in addressing poverty in the city, a lack of housing is still a key shortcoming at both city and state levels. In 2006 the State of São Paulo had a housing deficit of nearly 4 million units (Secretaria de Habitação 2013), defined both in terms of lack of housing (~1 million) and housing with substandard quality (~3 million). Created in 1949, the CDHU is the State owned company tasked with lowering this housing deficit. The CDHU is attached to São Paulo's Housing Department (Secretaria da Habitação), and functions by commissioning the construction of dwellings following tightly regulated parameters, and financing ownership through loans and subsidies for low-income population. Eighty per cent of the housing needs of the State fall within families earning between 1 and 3 times the legally defined minimum salary (CDHU 2011), equivalent to US$300 to US$900 per month, and broadly considered the level of monthly earning of low-income population. The large majority of the CDHU's dwellings are destined for families within this bracket (or 18 per cent of the State's households). The CDHU has been building in the order of 30,000 dwellings per year over the past decade (CDHU 2013), such that it is Brazil's largest housing agency. It operates with a substantial budget resulting from State level legislation directing 1 per cent of the State's VAT, known in Portuguese as the *Imposto sobre Circulação de Mercadorias e Prestação de Serviços*, to social housing; for the year 2012 it was expected that the State's social housing budget would be in the order of R$2 billion (€800 million). Whilst the geographical remit of the CDHU is the totality of the State, historically it has concentrated on building houses in 'the interior of the State' (that is the areas away from the coast) due to both the lower costs and higher availability of land. More recently, the CDHU has been directing its efforts to São Paulo's three metropolitan regions (Campinas, São Paulo and the Baixada Santista) in response to the more significant housing deficits in these regions of the state. This shift in geographical focus has also led to a change in the nature of the housing being built, from the low density, single storey housing familiar in the interior to higher density, residential towers (four to nine storeys).

At the same time as this evolution in the location and nature of social housing provision in São Paulo has taken place, there has been a growing interest in issues of sustainability as also central to the provision of housing in the region. This in turn is related to the growing importance of questions of housing quality within

São Paulo. While throughout the majority of the twentieth century the Brazilian census recorded only the presence or lack of housing, since the 1990s the census qualifies the question of social housing through the identification of substandard housing (e.g. some favela dwellers that, despite having access to housing, live under substandard conditions through a variety of shortcomings such as limited access to sanitary facilities). This opens up the possibility of approaching the housing problem not only from the perspective of housing shortage, but also of housing quality. The 'qualification' of the social housing debate represents a significant advancement in the social housing agenda, leading to questions on what is the appropriate quality to be provided and raising issues of 'housing with dignity', a demand first articulated by the city's social housing movements (Tatagiba *et al.* 2012). Sustainability and 'dignity' debates in social housing have come together towards an increase in housing quality through a series of building innovations that result in greater quality of life, such as natural lighting, construction materials of quality and, in the case of SHW, energy affordability.

Social housing and sustainability

When devising strategies for incorporating sustainability in social housing, the CDHU emphasizes the need to develop an approach that goes beyond environmental issues, placing a significant focus on interventions that also provide social benefits, in this way integrating the environmental, social and economic dimensions of social housing. In their view, as explained by a sustainability manager at the CDHU, 'there is no sustainability without social inclusion. There is no sustainability with people living in favelas. ... There is no sustainability with people living in precarious housing conditions' (Interview 1, State Government, São Paulo, October 2011). Drawing on this approach, the CDHU's sustainability actions focus on three dimensions (UNEP 2010b):

1 city retrofit via the incorporation of sustainable urban planning principles in existing neighbourhoods, so that social housing users also have easy access to transport and educational facilities, parks and other forms of social infrastructure;
2 the development of actions that promote social and economic sustainability at a community level, with a focus on creating capacity within beneficiaries through income-generating and other socio-organizational activities; and
3 the adoption of sustainability standards for buildings, including innovation with design standards as well as sustainable construction technologies.

The latter is addressed through a variety of means, with a focus on pilot experimentation in sustainable water technologies (such as shower timers, water saving devices and rain-water harvesting), and energy interventions (such as the SHW systems). Given the significant role that São Paulo's construction sector plays in GHG generation, the CDHU's experimentation with sustainable construction fits within the broad efforts of the State in responding to the challenges posed by climate change.

Both the State and the City of São Paulo have pioneered the development of policies and action plans in response to climate change. The city has been experimenting with different mechanisms for this purpose since 2005, when it became one of the first Brazilian cities to prepare a GHG emissions inventory (Prefeitura do Município do São Paulo 2005). Between 2005 and 2012, under the leadership of two successive mayors from Brazil's Social Democratic Party, São Paulo took a more active role in developing urban responses to climate change, partly driven by its involvement with the C40 Cities Climate Leadership Group. In 2009 the city of São Paulo instituted its Climate Change Policy (Law 14,933 of the 5th of June), becoming Brazil's first city to take this step. The policy explicitly recognizes the United Nations Framework Convention on Climate Change, while establishing an ambitious target for reducing 30 per cent of the city's emissions by 2012 against a 2005 baseline. This policy was followed in 2011 with a city-wide Climate Change Action Plan (Prefeitura do Município do São Paulo 2011), consolidating issues of climate change mitigation as an important local policy agenda. The Action Plan provides guidelines and priorities for future investment in the areas of transport, energy use, land use, building construction and waste resource management. Within this context, São Paulo's Solar Law (Law 14,459 of the 4th of June of 2007), mandating the use of SHW in all residences with more than 3 bathrooms, is seen as one of the main initiatives of the city in promoting low carbon. Similar climate change policies have been developed at the State level. In November 2009, the State of São Paulo launched its Climate Change Policy (Law 13,798 of the 9th of November), the first state level law of this type in Brazil, followed by the publication of the States' first Greenhouse Gas Emissions Inventory in 2011 (Governo do Estado de São Paulo 2011a). The policy includes specific guidelines requiring the construction sector to address issues of sustainable energy and energy efficiency in buildings.

The development of these policies and positions have enabled the City of São Paulo to play an important climate change leadership role within the global arena. In 2011 São Paulo hosted the C40 Mayors Summit, the biennial meeting of the Large Cities Climate Leadership Group. The meeting involved the mayors of Sydney, Addis Ababa and New York, alongside public representatives of nearly other 35 cities. The summit finished with a joint statement in which participating cities highlighted the role that megacities play in climate change, the need for these to have a voice in the debate, the importance of national governments and international treaties to empower the leadership of these cities, and the need to support them with the necessary resources to take action (C40 Cities 2011). This positioning reflects the ways in which climate change has come to be regarded as an issue which is central to the agenda of urban development in the city. As we discuss in detail below, this position has in part been shaped by the development of experimentation with solar hot water, and the ways in which this has enabled the merging of climate change concerns with those of providing essential services to those who are least able to afford them.

From experimentation to the mainstream: solar hot water and social housing

While the recent efforts by the City and State of São Paulo to place climate change on the urban policy agenda have highlighted the potential of solar energy, the CDHU has been experimenting with the potential of SHW systems for providing energy services in social housing since the early 2000s. SHW is a mature renewable technology for heating water for a variety of industrial and domestic purposes. Its domestic use is primarily aimed at heating water for basic household needs such as bathing, dishwashing, laundry and cooking, via collectors located in the roofs of the buildings (Figure 8.2). SHW uses heat transfer principles and materials to absorb radiant energy from the sun. As a domestic technology, SHW systems have been available in Brazil since the 1980s, when it was considered an expensive technology aimed at up-market neighbourhoods. However, successful pilot experiments in social housing were carried out by the CHDU between 2005 and 2007, paving the way for a more systematic attempt to mainstream the use of SHW in social housing. By 2011 the CDHU had installed over 35,000 SHW systems in both new and existing dwellings, using a variety of procurement modes and mechanisms aimed at the integration of this low carbon energy production mode into the construction and funding practices of the company. The motivations behind these efforts are multiple and overlapping, ranging from the need to systematically incorporate sustainability

FIGURE 8.2 Solar hot water heaters installed on a rooftop in São Paulo. Photo © A. Luque-Ayala, used with permission.

issues within the company's operations, to the endeavours of stakeholders within the country's electricity sector to smooth the country's peak electricity demand curve by limiting the use of electric showers. Together, they point towards a complex arrangements of practices of making, maintaining and living that transform SHW from an artefact into a tool directed towards the deployment of the will to improve, and the opening up of spaces of climate change experimentation in São Paulo, as explained in the next three sections. Making low carbon via SHW systems in São Paulo has meant problematizing energy in novel ways, aligning renewable energy technologies with social housing typologies, and reframing sustainability interventions in the context of social agendas.

Making space for solar hot water in São Paulo

The initial experimentation of the CDHU with SHW systems dates back to 2005, when a manufacturer of SHW systems approached the company with the purpose of establishing a pilot testing programme. Between 2005 and 2007, through site visits and interviews with the beneficiaries, the CDHU and the Brazilian Institute for Technological Research (IPT) monitored and documented the use of SHW in 50 detached individual dwellings. This first attempt to use SHW systems in social housing, in the municipality of Cafelandia located in the interior of the State of São Paulo, was considered to be a positive experience by all stakeholders involved. Whilst through this experience the CDHU was starting to develop its sustainability approach, users saw significant benefits thanks to the reduction in the electricity bill associated with replacing electric showers. This project bore all the hallmarks of a traditional demonstration project, designed to test both the attributes of the technology and the appetites of those involved. That the CDHU were able to develop their involvement with SHW systems from this initial base was a result of the conjunction of multiple energy problematics for which SHW appeared as a useful solution, as well as with the emergence within the organization itself of a new disposition towards housing provision, one in which issues of sustainability offered a means of ensuring that housing met certain standards of quality while also being provided at the scale required to meet the needs of the city's growing population.

Central to the emergence of SHW was the growing problematization of the provision of electricity, both nationally and in São Paulo. The historic development of hydroelectricity led to an abundant supply during the 1970s and 1980s which allowed for cheap electricity and the widespread use of electric technologies (Sternberg 1985), including water heating for showering purposes. Over 80 per cent of Brazilian dwellings use hot water for showering; 73 per cent use electricity for heating water and 99.6 of these use electric shower heads for this purpose (PROCEL/Eletrobras 2007), something that has significant implications for the stability of the national grid and its ability to satisfy demand. Whilst electric showers are highly efficient at the conversion of energy to hot water, they demand large amounts of electricity over short periods of time. As a result of the cultural practice of showering before the evening meal, Brazil's electricity peak demand curve spikes every weekday between 6pm and 8pm. Electric showers consume 8 per cent of the country's electricity production and are responsible for 18 per cent of the peak demand (Rodrigues and Matajs 2005).The steep daily peak demand curve associated with electric showers results in the need to continually invest in spare capacity in electricity infrastructure, and the presence of electricity generation and transmission assets that are underused during most of the day. In 2001, the challenges of providing for peak demand led to black-outs and the national imposition of energy rationing. Since the late 1990s and early 2000s, the National Agency of Electrical Energy (ANEEL), Brazil's electricity regulatory agency, has campaigned for energy efficiency and a long-term reduction in electricity use. In 2000, ANEEL pushed for regulations requiring private energy utility companies to spend 1 per cent of their net income on energy conservation (Law 9,991 of the 24th July 2000). Such investments are highly regulated and, since 2005 and since the pro-poor policies of the Lula government, a significant proportion of these funds have been earmarked for energy efficiency projects amongst low-income customers.

These pressures within the electricity system emerged in parallel to a second problematization of energy provision, reflected in the growing interest in the City and State of São Paulo in the issue of climate change. By coupling global concerns with local issues, climate change gains visibility as a realm of intervention. Despite the dominance of hydroelectricity in Brazil's electricity matrix, electricity use still makes a substantial contribution to GHG emissions in São Paulo. As of 2007, hydroelectricity accounted for 73 per cent of the total electricity capacity of the country and fossil fuels accounted for 16.5 per cent, whilst the rest was obtained via a combination of biomass, nuclear, wind and contracted imports. Out of the total emissions related to energy use in the municipality, 11 per cent correspond to electricity with the remaining 89 per cent correspond to the direct use of fossil fuels, largely towards transport uses. (Prefeitura do Município do São Paulo 2005). At the city level, the primary strategy for responding to climate change has been related to transport interventions. However, both state and municipal levels see the use of SHW systems as an important mechanism for responding to climate change, and São Paulo has become a national exemplar by becoming the first Brazilian city to pass legislation mandating the use of SHW in new construction (Law 14,459 of the 3rd July 2007). SHW systems, by offering a renewable energy solution that could reduce GHG emissions while also replacing electric showers and addressing peak demand, appeared to provide a solution for many of Brazil's electricity problems. SHW, as a malleable solution, was welcomed by many stakeholders, including the electricity industry who regarded them as a key means through which both to address peak demand and to deliver on their legal obligations for introducing energy efficiency measures for low-income populations.

In addition, changes within CDHU were also important in assembling a space for experimentation with SHW. Since the late 1990s, debates around sustainability had been slowly permeating the organization, starting with the instauration of the CDHU's first environmental programme in 1998. However, interviews carried out as part of the research leading to this chapter reveal how, during the early 2000s, the adoption of sustainability principles was largely associated with personal interests on the part of staff members rather than an outcome of an institutional approach. However, by 2007, and in the context of emerging discussions about climate change at State and City levels, the sustainability debate within the organization was mature enough for the establishment of a more systematic and institutional approach. At that time, the CDHU established a working group on the topic and created the position of energy efficiency manager, a high-level position with direct access to the company's executive director and the mandate of bridging the political, strategic and technical dimensions of implementing SHW systems within the organization. In this manner, CDHU sought to position social housing as a space within which solutions to the challenges facing energy systems could be found. For over a period of five years the CDHU increased its capacity to implement SHW through training members of staff, establishing partnerships with different organizations willing to donate SHW systems (particularly energy utility companies), developing a working relationship with the Brazilian SHW manufacturers association, putting in place programmes for quality control and development, and working together with academia in evaluating the impact of its strategy. Both climate change and the country's electricity constraints played an important role in legitimising this intervention. 'We're thinking in a different way, … always thinking about how not to generate carbon emissions', says the energy efficiency manager of the CDHU. But he also acknowledges how it was thanks to the contribution of SHW to overcoming the limitations of the country's electricity grid that 'the political will emerged, underpinned by those real factors … The political will to move ahead with solar hot water systems for low-income housing'.

Once the need to intervene through social housing to address energy problems had been established, and SHW systems regarded as the means through which this could be achieved, specific strategies and techniques needed to be developed in order to ensure that such solutions would be implemented. This meant locating available funding flows and translating the intention to implement SHW systems into pre-exiting procurement mechanisms, supply chains and highly-detailed contractual guidelines associated with each of the approximately 35 building typologies that the CDHU works with. SHW had to be calculated within existing protocols and working practices, made to fit within the system of providing and building social housing. It also meant developing working partnerships, engaging a range of actors previously unconnected to social housing provision and, through this, transferring the risks and responsibilities associated with this new form of experimentation. In this way, the SHW assemblage was expanded as the combined efforts of the CDHU and the material agencies of SHW systems needed to join forces with those of builders, architects, engineers, contractors as well as contracts, terms of reference, external audits, roofs, pipes, water meters, building structures and others in order to generate a new practice of energy production. Between 2008 and 2011 the CDHU used three different mechanisms to embed SHW systems into its institutional practices of social housing:

- *Building retrofit via SHW systems donated by energy utility companies.* This mechanism takes advantage of the national level legislation requiring electricity companies to invest in energy efficiency in low-income neighbourhoods. It relies on the donation of SHW systems, and is applicable only to existing inhabited buildings, as the utility company is required to certify to ANEEL a reduction in energy consumption in order to discharge its legal obligations. The key delivery agent and agent responsible for the purchase and installation of the SHW system is the electricity utility company that negotiates with the solar providers to install the device.
- *Direct purchase of equipment through public tenders.* In this mechanism, bidding solar manufacturing companies commit to supplying a large quantity of SHW systems at a fixed predetermined price. Through such a tendering process, the CDHU upgrades some of their units under construction or retrofits those that that they have built in recent years. It is therefore applicable to buildings that were originally conceived without SHW systems, but where the CDHU has identified a need or an opportunity for upgrade. The key delivery agent holding the responsibility for the SHW system is the SHW manufacturing company.
- *Incorporation of the SHW system within the normal construction practices of the CDHU.* In this final mechanism the housing unit is conceived and designed with SHW systems from the outset, and the equipment is considered as an essential house fixture (just like water faucets or any other integrated equipment), to be delivered by the building contractors responsible for each particular housing project. The key delivery agent and agent responsible for the SHW system is the construction company commissioned to build the housing units.

Whilst all these mechanisms share the same objective of introducing SHW in social housing, they rely on different procurement techniques and rationales. The first mechanism, where the key delivery actors are the commercial energy utility companies, relies on private intervention guided by a regulatory environment that requires a liberalized energy sector to invest in low-income population. The nature of the agreements signed between the CDHU and the electricity utility companies and the presence of mechanisms destined to measure success (in the form of external audits towards monitoring and reporting back on energy efficiency targets, a legal requirement established by ANEEL) opened a space for the final users to become an important part of the assemblage, as investments were not limited to the provision of SHW equipment, but included also equipment maintenance, energy awareness-raising activities, and user support. On the one hand, the CDHU insisted on the provision of an extended service on the part of the utility company, so that the latter's involvement would not be limited to the donation of SHW systems, but would also include a follow-up via user training and assistance in the case of failures. On the other hand, ANEEL's requirement of submitting third party verifications of energy savings to comply with the legal obligations imposed upon utility companies stimulated a concern with the success of the strategy beyond the donation of systems, which in turn meant engaging with users to ensure that they were using the SHW systems in the appropriate way.

In contrast, the second and third mechanisms function through the everyday logics of public investment and local government action: procurement mechanisms, public tenders and terms of reference are the key elements here. They play out in a space of market competition, with a variety of private sector companies (house builders and SHW manufacturers) competing for large-scale government contracts through lowering prices. In this case, the contractual terms of reference becomes the key instrument for rolling out SHW experiments in social housing. The contract, which details what is required and expected of the building contractor and/or manufacturer, is the reference point in case of disagreements in roles, responsibilities or expected qualities. Together, these mechanisms have enabled CDHU to scale up its experimentation with SHW systems from the initial 50 pilot systems in the municipality of Cafelandia of the mid-2000s to over 35,000 systems by 2011. The diversification of the experiment through different policy mechanisms corresponds to the need to adapt to a heterogeneous context of experimentation, without the need to extend the complex alliances between the core actors.

Maintaining the experiment: the challenges for the integration of water, energy and housing infrastructures

There is a clear move towards a widespread use of SHW in the majority of the operations of the CDHU. However, the company has not adopted a formal policy towards the use of SHW in *all* its dwellings. Whilst largely successful, São Paulo's experimentation with SHW has also been fragile, prone to break down and contested. For example: SHW manufacturers point to the inherent difficulties of incorporating their systems into pre-existing architectural typologies for social housing that did not consider SHW from the outset; the technical interface between SHW systems and other urban infrastructures or building systems has, on occasion, proved incompatible; and finally, the city and its densities impose limits on the widespread adoption of SHW systems. As a result, the experiment has undergone modifications, and it has been reworked in response to the need to ensure its upkeep. Yet persistent frictions amongst different urban infrastructures persist that transcend the assemblage of the experiment and in turn require processes of metabolic adjustment so that SHW can maintain its circulation within and through the city.

A significant aspect of maintaining the SHW experimentation undertaken by CDHU has been about re-imagining the energy/water interface within the home, particularly in the case of residential towers. Traditionally, energy and water infrastructures are autonomous and disconnected, such that the functioning of one does not affect the functioning of the other. SHW, in contrast, is a type of infrastructure where water and energy systems are combined and work in tandem. Maintaining the SHW experiment has therefore meant investing time and effort in solving the multiple problems that emerged out of this integration of systems and identifying the limits of the experiment: the dynamics of the water–energy interface meant that SHW became something that could only be installed in residential towers that had five stories or less. Central to these new interfaces have been practices of metering. Water metering has never been a common practice in Brazil. However, social housing tenants have increasingly requested water metering, particularly in residential towers, in order to address the social conflicts that arise from common billing methods that share the costs of water equally between households, no matter their level of use. With the arrival of SHW systems, the interface between individual water meters and the low carbon water heating technology proved problematic. In a limited number of cases, the lack of compatibility amongst cold (mains) and hot (solar) water has affected required water pressures and water supply. Thus, once technologies are assembled in the making of an experiment, further socio-technical innovation may be needed for their maintenance. In some sense, maintenance is a form a re-assemblage, but one that becomes visible only once the experiment has taken place.

However, the location of individual water meters at ground level, at the point where the municipal water pipe enters the building, precludes the possibility of using collective SHW systems on the roof of the building. This limits the overall capacity of solar hot water provided for the residential tower, since, instead of using a large hot water tank shared by all flats, each family has its own small tank (200 litres) on the roof. The resulting arrangement requires greater amounts of roof space per flat in order to supply the hot water requirements, limiting the size of residential towers where SHW systems are installed. The possible solutions for this hurdle (such as enabling the use of collective SHW systems by installing additional water meters at roof level of a quality standard capable of dealing with high water temperatures), which would increase the overall solar hot water capacity of the tower, increase the cost of the SHW strategy taking it outside of the realm of possibilities of social housing – a housing mode that needs to be affordable by definition. The insertion of SHW systems also created a lack of alignment between the metering and billing systems used by the water utility companies and the SHW strategy. In the words of an engineer at the CDHU:

> for each flat we would need two meters: one for cold water and one for hot water. But that metering is expensive, the equipment is expensive, and then the utility company would have to make two readings ... There are problems associated with the duplication of the individualized metering.

Ensuring that the SHW experiment can be maintained within the existing designs, fabrics and requirements of social housing has therefore been challenging, in turn reworking what it is that the SHW experiment can become.

In addition to adjusting SHW systems to the pre-existing water arrangements of the home, there has been a need to adjust the overall strategy of the experiment to the urban conditions of São Paulo, a process we have termed metabolic adjustment, where the experiment is affected by the physical or semiotic configurations of the city. Urban densities, dominant building typologies and the distance between the point of energy generation and the point of use play a role in enabling or disabling SHW in the city. In particular, the CDHU's strategy for the adoption of SHW has faced a new set of challenges as it moved from the interior of the State to the metropolitan centres, where higher urban densities and land values drive a change from extensive to intensive land use strategies. Whilst SHW systems in semi-detached, single family housing in the interior of the State have proved successful, the limitations and required adjustments associated with the use of SHW in residential towers have led to an approach that limits the use of SHW to towers that are less than five storeys high. First, as explained above, one way to overcome this difficulty is through the use of collective instead of individual SHW systems capable making a more efficient use of the limited roof space available, but technical and financial and constraints limit this. Second, urban density and the dominant building typologies of the city further challenge the use of SHW in social housing, by increasing the distance between the point of (solar) energy generation at roof level and its point of consumption in the form of solar hot water in the dwelling's bathroom. In residential towers, the distance between the SHW system and the user's shower, particularly for those in the lower floors, implies greater water wastage. The cold water that is in the pipes is discarded as the user waits for the hot water to arrive, pointing to a situation where a sustainability achievement in energy use could imply less sustainable forms of water consumption. In these cases, the waiting time between the moment the user opens the hot water tap and the moment hot water arrives varies from one to five minutes and more. As pointed out by a CDHU engineer, 'there is such a long time for the hot water to reach the downstairs flats [that] all that cold water is wasted' (Interview 2, State Government, São Paulo, November 2011). If the wait is too long the user no longer relies on the SHW system and switches back to the use of electric showers, rendering the SHW system useless. These two challenges affect primarily those urban areas with a prevalence of high rise towers, a common condition within social housing. It is a limitation with important implications for two reasons: on the one hand, the CDHU's response to housing shortages in highly urbanized areas is through vertical development, in order to maximize land use and minimize the investment associated with land purchases; on the other, this type of response is most common in São Paulo's three metropolitan regions, characterized by expensive land prices and a significant housing shortage. In summary, maintaining metabolism circulations is closely related to the need for reconfiguring both the experiment and the house, to adjust the experiment, in a sense, to reassemble it, and to adapt the existing urban infrastructure to the disturbances created by the experiment's insertion. But, most of all, this experiment shows the need to readjust a whole bunch of discourses of housing, energy and water provision, including the systems boundaries and their operation, which is to have a long-lasting influence both with regard to housing provision and the possibility of launching further experiments. Experimentation with low carbon is not only about using new technologies as part of the roll out of sustainable construction initiatives, but also about understanding how these technologies interact over time with pre-existing technological arrangements as well as final users.

Bringing SHW to life: embodying dignified housing

While significant benefits in relation to climate change and the optimization of Brazil's electricity grid proved to be significant in the making of SHW as an experiment, the critical dynamic that has animated the extension and consolidation of SHW systems within the CDHU has been their ability to offer a tangible demonstration of how social housing can contribute to improving the quality of life of its tenants through reducing the costs of living over the lifetime of their residency. Rather than being a matter of the provision of housing per se, CDHU have sought to re-orientate the debate to one in which 'housing with dignity' is at the heart of its service delivery. The need for 'housing with dignity' has been a longstanding claim of São Paulo's social housing movement which, through this debate, has configured itself as a political actor demanding not only a 'right to the city' (embodied in a dwelling in a central location) but also making additional claims around the quality of such dwelling (Tatagiba *et al.* 2012). What is at stake is a set of values subsumed in the notion of 'housing with dignity'. Yet 'dignity' represents more than quality housing; it is also an attempt to subvert the humiliation that accompanies a condition of vulnerability (Kowarick 2009). Social housing programmes in São Paulo have responded to this debate by enshrining notions of 'housing with dignity' within the state's Housing Plan 2011–2023. This establishes 'the continuous search for the promotion of dignified housing, which ensures the qualities and conditions of habitability to all individuals in an egalitarian manner' as one of the foundational principles of the State's housing policy (Governo do Estado São Paulo 2011b: 235; translated from Portuguese). Under this logic, 'quality' housing is associated with providing 'dignified' housing, or housing through which it is possible to lead a dignified life; SHW systems, through the benefits they generate in lowering living costs and increasing energy affordability, become an important element in the delivery of 'housing with dignity'. The ability of SHW to demonstrate how this works in practice, providing a real demonstration of the benefits of such an approach, together with its role in creating new forms of subjectivities about what tenants might come to expect from social housing, have been critical in establishing SHW as a form of experimentation that is able to move into mainstream agendas in the city.

Established social housing strategies are particularly sensitive to the adoption of new practices or technologies that could increase the price of housing units. For this reason, the decision of the CDHU to experiment with SHW systems was not to be taken lightly, as the addition of new technologies to the dwelling would upset a carefully-balanced equation between the amenities provided, the cost of each unit, and the number of units built per year. Increasing the dwelling cost could have a negative impact on the total amount of population served per year, a sensitive issue in a highly-political local context, where well-established social housing movements vocally advocate for a reduction in the housing shortage, and politicians are judged on their ability to deliver social benefits.

The adoption of SHW systems imposes pressures at the core of the tension between quantity and quality in housing infrastructure. The sustainability imperative became an integral element of the 'housing with dignity' discourse, enabling a strategy that, through an increase in technology, would inevitably result in higher costs and therefore pose an impact on the company's capacity for delivery (when measured against the amount of units built). Through multi-stakeholder engagements and discussions, the CDHU qualified the meaning and nature of sustainability in social housing, and the SHW system became one of several sustainability interventions contributing to housing with dignity and a better quality of life: a wide range of amenities and facilities; a house built with quality materials; better air circulation enhancing indoor air quality; and illumination strategies reliant on natural light. In the case of SHW systems, the confluence between the sustainability and housing with dignity debates also relies on the long-term benefits for the final users and for society at large. These are achieved via lower electricity bills (up to 30 per cent savings) and through the promotion of a set of values that encourage environmental conservation. In this case, the 'lifecycle' logic of sustainability experiments in housing, where the final user gets the benefits by a reduction in housing costs over the life of the building (Lovell 2004), is applied in the context of low-income users. Social housing users in Brazil are estimated to spend between 20 and 24 per cent of their energy bill on water heating (UNEP 2010a), and a large proportion of this can be saved through the use of SHW. Such economic benefits meaningfully contribute to poverty alleviation, as explained by a user of SHW systems living in a CDHU project in the municipality of Cubatao, Greater São Paulo:

> We live here 4 people. We like the SHW system. First of all because of the economy. But also because it helps with environmental issues. We moved here 3 months ago. Our energy bills have been as follows: the 1st month R$19, the 2nd month R$R28, and the 3rd month R$38. [...] before moving into this house, we were paying between R$70 and R$90 because of the electric shower. We were using an electric shower, but since we moved in to the new house we have only been using the SHW. If we were to use the electric shower the energy bill would be higher.
>
> *(Interview 3, Resident, Cubatao, November 2011)*

For beneficiaries of social housing, moving into their new home often means a series of additional financial burdens, on occasions beyond their ability to pay. The regularization of housing conditions implies, for the first time in many cases, the need to pay water and electricity bills. By reducing the expenditure on utility services, the SHW system frees scarce financial resources that can be made available for other needs or commitments associated with housing provision, such as the mortgage and the condominium fees. Thus, the financial benefits associated with SHW systems play a role in managing financial risks for all parties involved, including the CDHU. As a quality manager at the CDHU explains, the SHW systems provide important financial benefits for the user in the context of a new set of financial responsibilities associated with formal housing:

> Why the solar hot water? There is an environmental aspect, but it is a system that generates lower costs … We work with … people of low income. Therefore you are generating possibilities for the maintenance of their house. Because very often people who come from a background of informal housing do not pay electricity, do not pay water … Because they are doing their own connections … And in the new model they have to start paying condominium fees, so that is a problem.
>
> *(Interview 4, State Government, São Paulo, October 2011)*

In this way, and through a low carbon energy strategy, the CDHU managed to enter the sustainable construction debate with a perspective that makes a significant emphasis on issues of inclusion, dignity and social justice. The incorporation of building sustainability features in social housing thus acts as a complement to the social agenda that the company is already pursuing via the provision of housing for low-income population. The example shows how the creation of low carbon subjects is created through the alignment of climate change experimentation with the immediate needs of residents in the houses provided with SHW: by demonstrating how an environmental benefit may ultimately impact on the local residents' personal budgets there is a separation of interests – individual versus collective – and a subsequent conflation oriented towards the enrolment of private interests in the collective project.

Impacts and implications

The use of SHW in São Paulo's social housing has given rise to a new politics of low carbon energy production. It is a politics based on the integration of urban infrastructures at the point of use, specifically the dwelling. By replacing high-demand electricity-based systems, such as the electric shower, this intervention has contributed to the optimization of the country's electricity grid and, in doing so, to a reduction in carbon emissions. Establishing SHW systems required considerable work, such that, beyond policy documents and political statements, a multiplicity of small agencies are required in the making, maintaining and living of low carbon. These agencies are grouped as an assemblage, where fragility, contingency and dispersion play a key role in the final outcome. The agencies are both human and non-human, material and semiotic, as success depends on policy makers and project managers as much as it does on engineers, architects, technicians, builders, contractors, contracts, terms of reference, quality standards, discourses, urban densities, building typologies, and finally, users. As the experimentation with SHW in São Paulo has come to be related to mainstream agendas for urban development and the provision of social housing, three critical dynamics have been central: the positioning of SHW as an appropriate urban response to climate change and energy issues in social housing, through a transit from conceptual and physical peripheries to the central spaces of decision-making; the contribution to social housing strategies at national level, sparking further processes of scaling up the use of SHW; and the positive but sometimes conflicting impact of the process of experimentation for promoting locally-grown green industries.

SHW systems: moving from periphery to centre

In 2005 the CDHU initiated a pilot experiment with 50 SHW systems in the municipality of Cafelandia. Six years later the company had installed over 35,000 systems, with thousands more in the pipeline. What was once a pilot initiative became a mainstream policy, with the CDHU installing SHW systems where it deemed this technically feasible. Experimentation with SHW started in the periphery – the lower density interior of the State – and progressively moved towards the higher density metropolitan areas, including the city of São Paulo. As the experiment reaches the city of São Paulo, a greater political impact is expected, given the political clout that projects in the city carry with respect to the regional and national political context. This is likely to increase the visibility of the strategy, provide new technical insights and valuable publicity towards a further legitimization of the intervention. As a staff member of the CDHU puts it: 'everything that you do in the city of São Paulo gains greater visibility ... [and] the more people aware of how the CDHU is shifting towards a policy of sustainability, the better' (Interview 6, State Government, São Paulo, October 2011). In this way, the strategic mobilization of the experiment from the periphery to the core forms part of the means through which it has gained traction. The shift from periphery to centre is not only spatial: it is a shift that also mirrors the growing popularity of the system within the organization, and a gradual progression and learning curve in the adoption of technology. The SHW system moves from the periphery of the discourse and practice to the centre, as it also moves towards greater socio-technical complexity and broader challenges in relation to the integration of housing, water and energy infrastructures. Both the geographical move from periphery to centre, as well as the transition from lesser to greater complexity, are highly related to specific urban conditions: land use models and the resulting urban densities and housing typologies.

Getting the experiment to travel: scaling up SHW in social housing at the national level

The experience of the CDHU with SHW systems is providing a template for the Brazilian national government to upscale the use of SHW in social housing all over the country. In 2009, the Brazilian federal government launched 'Minha Casa, Minha Vida', a large-scale social housing programme aimed at scaling up social housing provision in Brazil. The programme counted on an initial investment of R$34 billion, aiming for 1 million homes during Phase 1 (2009–2010). An additional 2 million homes are planned for Phase 2, scheduled for 2011–2014. The programme is run by the Caixa Econômica Federal, the largest government-owned bank in Latin America. Building on the experience of the CDHU, the Minha Casa, Minha Vida programme has decided to include SHW systems in its dwellings. In

Phase 1, approximately 41,500 housing units included SHW systems. In Phase 2, all dwellings for families earning between 1 and 3 minimum salaries will include SHW systems, except those to be located in residential towers. It is expected that approximately 380,000 homes in Phase 2 will have SHW systems. The emphasis on families earning between 1 and 3 minimum salaries is related to the housing funding mode, where dwellings in this population group are fully financed and largely subsidized by the Caixa Econômica Federal, whilst those of families earning over 3 minimum salaries respond to market mechanisms and therefore the user would have to fund the technology. The programme has limited the available funding for each SHW system to R$2,000 (€780), including installation costs.

The terms of reference, quality guidelines and procurement mechanisms developed by the CDHU are providing a template for the federal government to move forward in the operationalization of SHW in social housing. Staff members of the CDHU, who gained extensive knowledge on SHW systems through the organization's experimentation process, were invited to join a variety of committees aimed at shaping the sustainability policies of Minha Casa, Minha Vida. However, the experiment travelled not only via the experience of those involved, but also through calculation devices, primarily through the terms of reference originally developed by the CDHU to regulate relationships with those responsible for supplying and installing SHW (e.g. energy utility companies and SHW manufacturers). This exchange of information, in the words of the energy efficiency manager of the CDHU, allowed Minha Casa, Minha Vida to 'gain time'. He explains how, through this process of learning from previous experimentation, the Minha Casa, Minha Vida programme will manage to avoid experiencing 'the difficulties that we [the CDHU] went through. They are going to start in a more mature way ... There is a working group in Brasilia that meets for this topic only. I am part of that group, [where] I explain the things I know' (Interview 5, State Government, São Paulo, November 2011). These terms of reference, functioning as contractual devices governing relationships, embody a multiplicity of forms of knowledge capable of travelling, such as:

- *Finance modes and procurement mechanisms* – establishing the conduits and mechanisms associated with the delivery of SHW.
- *Design templates* – determining optimal sizes and physical interfaces between different energy, housing and water infrastructures.
- *Standards and certification systems* – such as quality standards.
- *Distribution of responsibilities* – establishing, for example, who is responsible in case of break down.

Finally, the logics that frame the use of SHW in social housing, particularly an understanding of SHW as a combined response to the challenges posed by climate change and the country's peak electricity demand patterns, are also embraced by Minha Casa, Minha Vida. This is evidenced by the promotional material developed by the programme, which sees SHW as 'a clean energy source that contributes to a reduction in GHG emissions, in line with the National Climate Change Plan, whilst also delaying the construction of new facilities for the generation and distribution of electricity' (Governo Federal do Brasil n.d.: 2; translated from Portuguese).

The CDHU was consulted within the programme Minha Casa, Minha Vida because of their varied experience, including experimentation with SHW. There was an exchange of personnel, and the person in charge of energy projects at the CDHU (in São Paulo) joined the committees that were planning Minha Casa, Minha Vida (in Brasilia) in order to share experience. The CDHU also shared key documents with Minha Casa, Minha Vida, such as the terms of reference associated with the strategy to purchase SHW systems. In doing so, the experiment facilitated the circulation of ideas beyond the context of São Paulo. The experience in São Paulo did not only generated expectations and new spaces of low carbon possibilities, but also a particular ethos that integrated ideas of housing with dignity at the core of housing practices. But we do not have evidence of the extent to which values and practices also travel, together with the ideas about technology operation and implementation.

Promoting the development of a green industry

Throughout its experimentation with SHW, the CDHU has been aware of the power of the public sector to influence the development of an emerging 'green' industry. The decision of the CDHU to install thousands of SHW systems in its housing stock facilitates economies of scale, capable of lowering the cost of low carbon technologies while making them more competitive. The significant role that the CDHU plays in the development of a SHW market is acknowledged by a sustainability manager working with the organization: 'We pay and buy in large scale, and in that way we create a market' (Interview 1, State Government, São Paulo, October 2011). The CDHU has used this power, through negotiations with the Brazilian association of SHW manufacturers, to push for technological development, quality improvements, and a broader recognition of additional sustainability dimensions, such as the manufacturer's extended responsibility for product disposal after its life cycle. The experiment has created a market, has strengthened industrial capacity, and thus it has shown how the state can intervene to open the market. As in Monterrey or Bangalore, bringing the experiment from an assemblage to an intervention that is truly lived also required building a product that those who live the experiment could 'buy into'. In the wider context of global capitalist circulations, that also means creating a market for the circulation of materials and capital flows, together with the creation of subjects willing and prepared to intervene in such markets. This is not just the creation of consumers for SHW in São Paulo, but rather the creation of a particular type of subject who may consume the technology in different ways (through using it directly, through installing it) but who, in any case, would buy into the broader conception of the technology and its construction as a response to a given problem, as a key instrument to operationalize the will to improve. Crucially, the state is portrayed as an active agent who can, indeed, manage these circulations and control technology, but, as the case shows, this is not done by the state alone, but by a complex systems of alliances and rationalities.

The interaction between the CDHU and SHW manufacturers provides an example of low carbon technological innovation and development promoted and governed by state intervention. This has been achieved through a dual logic of driving the SHW industry towards self-regulation and control (primarily via quality standards), while tightening the terms of reference and legal contracts guiding the interaction between both parties. However, this process of mutual adjustment has resulted in stress in the relationship between the CDHU and the SHW manufacturers. The CDHU claims that the gradual change and inclusion of greater levels of detail in the terms of reference governing the interaction between them and manufacturers reflects the need to clarify responsibilities, specifications, ensure product quality and adjust the systems to the specific housing typologies in use. Through the use of terms of reference, staff members at the CDHU claim that the organization has been able to close legal loopholes previously allowing manufacturers to lower the quality of the products, whilst also pushing for better energy performance and an increase in the obligations expected from manufacturers. On the other hand, manufacturers interviewed feel that the work with the CDHU has been a not very easy process of learning by doing, where many of the financial costs associated with the inevitable mistakes and failures of the experimentation have been borne by the manufacturing sector.

The use of SHW systems in social housing provides an example of how a low carbon strategy is reinterpreted within the domain of the social. Here the key aspect that provides legitimacy is not the extent to which the experiment results in low carbon or energy efficiency infrastructure, but the ways by which these logics are mobilized in support of broader political objectives, such as those associated with housing provision and, through that, poverty alleviation. But the SHW experiment in São Paulo does not operate exclusively in the space of social agendas. The SHW systems perform an important market function by freeing the energy market from the constraints imposed by the daily peak electricity demand. This enables utility companies to devise more flexible strategies, and frees up financial resources originally earmarked for the expansion of the electricity network in order to cope with such daily demand in the context of supply limitations, opening up new opportunities for the circulation and marketization of energy. The material water–city–energy interface plays an important role in enabling or disabling this experimentation with SHW, which, despite the creation of new marketization opportunities, enacts social logics through supporting issues of social justice in the city in relation to climate change and social housing.

References

Anderson, B. and McFarlane, C. (2011) 'Assemblage and geography'. *Area*, 43: 124–7.

Bennett, J. (2010) *Vibrant Matter: A political ecology of things*. Durham, NC: Duke University Press.

C40 Cities (2011) Press release: C40 São Paulo Summit letter to Rio + 20. United Nations Conference on Sustainable Development. Available online at: http://www.prefeitura.sp.gov.br/cidade/secretarias/desenvolvimento_urbano/comite_do_clima/c40/en/press/index.php?p=49005 (accessed 14 July 2014).

CDHU (2011) *Sustainability Report 2011.* São Paulo: Governo do Estado de São Paulo.

CDHU (2013) *Distribuição Territorial da Produção.* Available online at: http://www.cdhu.sp.gov.br/producao-new/distribuicao-producao.asp# (accessed 30 June 2014).

Governo do Estado de São Paulo (2011a) *1° Inventário de Emissões Antrópicas de Gases de Efeito Estufa Diretos e Indiretos do Estado de São Paulo: Período 1990 a 2008.* São Paulo: Secretaria do Meio Ambiente/CETESB.

——(2011b) *Plano Estadual de Habitação de São Paulo.* São Paulo: Secretaria de Habitação.

Governo Federal do Brasil (n.d.) *Sistema de aquecimento solar de água no programa Minha Casa Minha Vida.* Brasilia: Ministério do Meio Ambiente.

Kowarick, L. (2009) *Viver em risco*, São Paulo: Editora 34.

Lovell, H. (2004) 'Framing sustainable housing as a solution to climate change'. *Journal of Environmental Policy & Planning*, 6: 35–55.

Prefeitura do Município do São Paulo (2005) *Inventário de Emissões de Gases de Efeito Estufa do Município do São Paulo.* São Paulo: Centro de Estudos Integrados sobre Meio Ambiente e Mudanças Climáticas (Centro Clima).

——(2011) *Diretrizes para o Plano de Ação da Cidade de São Paulo para Mitigação e Adaptação às Mudanças Climáticas.* São Paulo: Comitê Municipal de Mudança do Clima e Ecoeconomia.

PROCEL/Eletrobras (2007) *Avaliação do Mercado de Eficiência Energética no Brasil.* Pesquisas de Posse de Equipamentos e Hábitos de Uso Rio de Janeiro: Eletrobras.

Rodrigues, D. and Matajs, R. (2005) *Brazil Finds Its Place in the Sun: Solar water heating and sustainable energy.* São Paulo: Vitae Civilis.

Secretaria de Habitação (2013) *Conheça melhor a Secretaria de Habitação.* Available online at: http://www.habitacao.sp.gov.br/secretariahabitacao/conheca_melhor_a_secretaria_de_habitacao.aspx (accessed 30 June 2014).

Shove, E. and Walker, G. (2007) 'Caution! Transitions ahead: politics, practice and transition management'. *Environment and planning A*, 39: 763–70.

Sternberg, R. (1985) 'Large scale hydroelectric projects and Brazilian politics'. *Revista Geográfica*: 29–44.

Tatagiba, L., Paterniani, S. Z. and Trindade, T. A. (2012) 'Ocupar, reivindicar, participar: sobre o repertório de ação do movimento de moradia de São Paulo'. *Opinião Pública*, 18: 399–426.

UN-Habitat (2010) *São Paulo: A tale of two cities.* Nairobi: UN-Habitat/SEADE – Fundação Sistema Estadual de Análise de Dados.

UNEP (2010a) *Assessment of Market-available Technologies and Solutions to Improve Energy Efficiency and Rational Use of Water in Social Housing Projects in Brazil.* Paris: UNEP/SUSHI/CBCS/CDHU.

——(2010b) *Lessons Learned: Sustainable solutions in social housing from the experience of the Housing and Urban Development Company of the State of São Paulo.* Paris: UNEP/SUSHI/CBCS/CDHU.

9

CLIMATEERS – PIONEERING A NEW RESPONSE TO CLIMATE CHANGE IN HONG KONG?

Introduction

Hong Kong, referred to colloquially as the 'pearl of the orient', stands out as one of the dense nodes of light that mark China's coastal cities on NASA's now-familiar image of the earth at night. Situated in the Pearl River Delta, the Hong Kong Special Administrative Region (SAR) passed from British governance in 1987 to form a semi-autonomous part of the Chinese state. A thriving commercial and business centre, the governance of Hong Kong has been characterized as 'a top-down central government with a neoliberal market environment that fosters a powerful business sector' (Chu and Schroeder 2010: 289). Within this context, the local state has focused on economic growth rather than sustainability or ecological modernization (Gouldson *et al.* 2008). However, during the 2000s, as the Chinese central government sought to develop a low carbon economy in response to climate change while business interests and non-state actors in the city simultaneously began to mobilize for action, there has been a gradual shift in this position. During this period, the Hong Kong government began to accumulate a 'significant body of scientific, technological and economic climate change knowledge by commissioning a series of consultancy reports' (Francesch-Huidobro 2012: 1), and initial discussions on the relevance of addressing climate change were held in the Legislative Council (Chu and Schroeder 2010). In 2008, an Inter-departmental Working Group on Climate Change was established, and in 2010 the Environment Bureau published a consultation document setting out, for the first time, the actions required within Hong Kong. Acknowledging the scale and importance of the challenge, the report identified changes in the energy supply mix that powers Hong Kong, proposing to address it primarily by importing nuclear electricity from new power stations in China. At the time of writing, no further changes have been made on that position.

The position adopted by the Hong Kong government sits in contrast to that of the emerging coalition of private interests in the city – including multinational corporations, environmental NGOs, development organizations and community groups – who have argued that climate change action should involve changing the way in which energy is used in the city rather than merely increasing supply. The focus of this coalition reflects concern about the potential environmental implications of nuclear power as well as an emerging sense of moral responsibility to reduce the city's energy consumption. Sitting within China, with its non-Annex 1 status in the terms of the international climate change conventions, Hong Kong is not obligated to reduce greenhouse gas (GHG) emissions. However, for this emerging coalition, the levels of energy consumption and economic development within the city are such that Annex 1-comparable targets and timetables for addressing GHG emissions should be pursued. At the same time, in a city whose electricity consumption is literally on display – the evening light show is an important tourist attraction – the idea of a low carbon transition poses a considerable social and cultural challenge.

In this chapter, we examine how, in the context of a relative policy vacuum, this coalition of private actors began to mobilize to govern climate change in the city through a series of experiments which sought to target dispositions towards energy use. In the next section, we detail the emergence of this coalition. After that, we turn our attention to one particular initiative, the Climateers programme, developed in partnership between WWF and HSBC: we examine the working of this experiment in practice, through the processes of making, maintaining and living. The final section considers some of the central implications this case poses for how we understand urban climate governance as an art of experimentation.

Private authority and climate change governance in Hong Kong: creating a new agenda?

In 2007, climate change rose rapidly up the policy agenda in Hong Kong, driven by an emerging coalition between business and non-governmental organizations in the city, the growing interest from the Chinese government towards the issue and the increasing attention given to the issue globally. While broader environmental concerns, including air and water pollution, had long been of concern to the Hong Kong SAR's Government, climate change had yet to be considered as a matter for policy intervention in its own right. For those in the city seeking to mobilize around climate change, the initial issue was that of making climate change into a policy problem:

> without having mapped it out in incredible detail … we [realized] that because this issue had appeared so rapidly we would certainly have to focus on the kind of public piece first; getting a bit more awareness. … We very much realised that the Hong Kong government is a follower and not a leader,

> so we were clearly going to get general awareness in society then work with those most progressive elements of business on the solution side and then … And then … get involved in the policy discussion when the government was willing at least to talk about it.
>
> *(Interview 1, NGO, Hong Kong, July 2011)*

Central to this ambition was the creation of constituencies that would signal the significance of climate change as a matter of concern and that could be mobilized in support of action. These constituencies were broadly seen to include 'the public' on the one hand and the 'most progressive elements of business' on the other, and were seen as being critical to both ensuring support for a policy response to climate change and to the workings of any such response. Some 67 per cent of Hon Kong's GHG emissions in 2008 were generated through the use of electricity, 90 per cent of which was used in buildings. A further 18 per cent of the city's emissions came from transportation (HKSAR 2010). Changing the generation and use of electricity, particularly in the built environment (offices and homes), therefore came to be understood as key to climate change mitigation by the loose coalition of business and environmental organizations who sought to increase the profile of climate change between 2007 and 2011.

Seeking to mobilize climate change as a policy problem also required a means through which to demonstrate that such constituents were both willing and able to take action. Rather than being a problem in need of governmental intervention, climate change was positioned as a critical issue to address, and one for which solutions were available. This framing of climate change spoke to the tenor of the debate in the late 2000s, where a range of novel forms of carbon markets, public–private partnerships and voluntary schemes were being developed in response to growing international concern (Hoffmann 2011). The intervention of international organizations such as HSBC, WWF, FoE, C40, Oxfam and The Climate Group in Hong Kong, alongside the growing interest amongst the business community and NGOs in the city, explains how a strong coalition of the willing between the business and environmental communities was forged. This alignment of multiple actors has been key in developing the 'will to improve' climate change. Together with the range of forums, reports and events that were formed during this period, and which served as a means of building the coalition required to garner political interest, a range of specific programmes and schemes were devised that could demonstrate the potential of responding to climate change. These initiatives specifically targeted the issue of energy consumption in the built environment, and included a range of measures from the development of new voluntary standards for green buildings, the implementation of new energy management systems within businesses, events such as 'Earth Hour' and 'No air conditioning night' and the 'Dim It' campaign designed to draw attention to the over consumption of electricity, and schemes which sought to engage businesses and households in measures to reduce their everyday demand for energy (Francesch-Huidobro 2012; Chu and Schroeder 2010). In the next section, we

FIGURE 9.1 Advertising the Friends of the Earth PowerSmart Contest at the Ferry Terminal, Hong Kong. Photo © H. Bulkeley.

consider how, within this context, Climateers emerged as a key experiment that sought to intervene in the reconfiguration of carbon governance in Hong Kong.

Social innovation for governing climate change: the case of Climateers, Hong Kong

> The Climateers programme encompasses both the web portal and a wide range of supporting offline-community activities and on-going projects. By promoting climate change awareness of the general public and strategic communities, Climateers aims to create a hub for a movement that will put Hong Kong on the map with other world cities on a common mission to tackle climate change. As a hub for climate change information, solutions, networking and social & cultural discussion, backed by a cutting edge carbon calculator specific to a Hong Kong lifestyle, www.climateers.org is a first-of-its kind initiative in Asia.
>
> *(WWF Hong Kong 2013a)*

> what we call climateers – it's volunteers and pioneers in climate change. So they are the early adopters, the first group to respond.
>
> *(Interview 2, NGO, Hong Kong, July 2011)*

Climateers was initiated in 2007 by WWF, with funding of HK$4.45 million from HSBC (HSBC 2013), as a programme aimed at establishing both a 'hub' for public responses to climate change, and a group of early adopters who would be able to calculate and reduce their carbon footprints. The scheme was born out of a long-standing partnership between the two organizations, first established in 2002 as part of HSBC's five year Investing in Nature programme, and developed through the HSBC Climate Change Partnership (2005–2010), an international programme developed by HSBC with The Climate Group, Earthwatch, WWF and the Smithsonian Tropical Research Institute, with an aim to 'inspire action on climate change' internationally as part of HSBC's commitment to addressing climate change both within and beyond its business operations (The Climate Group 2013).

Central to the development of the first phase of the programme was the creation of the first online carbon calculator in Hong Kong, a web portal which enabled Climateers to compare their carbon footprints and provided advice on creating a low carbon lifestyle, as well as a competition for the public to devise ideas that would contribute to low carbon living (Figure 9.1). More than 15,000 people became Climateers, of which '5,855 of them have been using the carbon calculator to track their personal carbon footprint since 2007' including '270 HSBC colleagues' (HSBC 2013). In addition, a specific initiative, Climateers Ambassadors, was devised as a means of engaging with youth communities in Hong Kong. This scheme was developed through a partnership with the Hong Kong Junior Police Corps and the Hong Kong Federation of Youth Groups – two organizations which WWF had previous experience of working with. It provided a means for WWF to test the development of the ideas behind the programme and the workings of the website, as well as trialling the potential for a more intensive programme of engagement. Climateers Ambassadors not only had access to the carbon calculator and general information as provided to the public through the website, but were also given a series of workshops held at WWF's wetland (Mai Po) and marine (Ho Hai Wan) nature reserves. The Ambassadors programme was then opened to the wider public. Heavily oversubscribed, some 100 participants were selected to follow the same experiential learning programme. For those who became Climateers Ambassadors, 'according to their self-declared data … they [together have made] … a 25 per cent reduction in their personal [carbon footprint]' (Interview 2, NGO, Hong Kong, July 2011).

The end of 2010 marked the completion of the first phase of the Climateers programme and the culmination of the financial support received by WWF from HSBC, as the HSBC Climate Change Partnership was completed and funding was sought from the Environment Protection Department. This second phase of Climateers marked a change in direction (Table 9.1). While the web portal, carbon calculator and information on low carbon living continued to be maintained and developed, the Climateers Ambassadors scheme was substantially revised to focus on two area-based communities – Wan Chai and Sai Kung. This shift in the nature, focus and partnerships involved in Climateers served to further embed the programme in the city, and at the same time raised a host of new challenges in terms

TABLE 9.1 The two main phases of the Climateers programme

	Phase I	*Phase II*
Initiators	WWF and HSBC	WWF
Partners	Youth groups	Community groups
Funding	HSBC	EPD
Problematisation	Need to reduce energy demand in the built environment and increase levels of public education on climate change	Need to transform daily practices towards low carbon alternatives in order to reduce energy demand
Innovation and experimentation	• Carbon calculator • Experiential learning programme	• Carbon calculator app • Community-based action programme
Tactics and techniques	• Competition • Immersion • Carbon calculation • Viral marketing • Peer-to-peer learning • Information/education • Demonstration	• Competition • Immersion • Carbon calculation • Viral marketing • Peer-to-peer learning • Information/education • Demonstration

of seeking to develop and implement low carbon living in the city. In the case of Climateers, we find that each aspect of experimentation – its making, maintaining, and living – has been important in sustaining social innovation, operating in tandem with the problematization of climate change and the alignment of interests and techniques that enabled the intervention to take place, simultaneously set in train the means through which the programme can be maintained and require the uptake of new forms of low carbon conduct. In this case, we find that the process of experimentation is not a 'one off'; rather, as we explore in detail below, Climateers is reinvented through a new alignment of actors and interests, and the development of new forms of social innovation, even while it retains the same rationale and techniques in order to govern climate change as behaviour change in the city.

Making: assembling Climateers at the intersection of private interests

In the absence of the engagement of the government of Hong Kong with climate change, the emergence of the issue on the agenda came to be driven by a range of private actors who were concerned to make it public: both in order to garner support for moving climate change up the policy agenda but also in terms of framing climate change as an issue which required a public response. Through the coalition of business and environmental organizations who became involved in this political space, climate change came to be understood as a problem of managing energy demand, particularly in the built environment. Reporting a study of stakeholder views on how climate change should be addressed in 2010, The Climate Group found that:

> … most interviewed stakeholders see energy efficiency as a low-hanging fruit and a quick-start to mitigate climate change. While in China buildings consume around 18% of the country's total energy, 89% of the electricity is being consumed by buildings in Hong Kong.
>
> *(The Climate Group 2010: 6)*

This in turn has meant that a range of elements that serve to make up energy demand (from the building fabric itself, to the regulations guiding its development, the management systems used to control energy in buildings, the use of air conditioning, office cultures, and everyday practices of cooking, washing and lighting) have come to be entrained within the notion of responding to climate change. Though these are often amalgamated, different tactics place emphasis either on the need to intervene in material terms, in the form of new technologies and systems, or on the importance of engaging and sustaining a behavioural response, of achieving a shift in how much energy is consumed by employees and residents. For the partners in the Climateers programme, the idea that addressing climate change was a matter of engaging the public and changing their behaviour was central. HSBC, through the HSBC Climate Change Partnership, had committed to engaging its employees with a programme of learning and action on how to address climate change at work and at home. WWF had advanced the concept of the ecological footprint internationally during the 1990s and 2000s. In 2007, they were first approached by businesses in Hong Kong concerned to understand what carbon reporting involved, following a recommendation from the Carbon Disclosure Project that they should report their carbon emissions publicly. In the commercial sector, WWF then established the Low Carbon Manufacturing Programme and the Low Carbon Office Operations Programme to work with businesses in Hong Kong to address their carbon footprints. At the household level, seeking to engage individuals with behavioural change also addressed a long-standing concern about how NGOs build engagement:

> I think it's always difficult to … get their participation, because we are competing with many other activities, you know in Hong Kong. You know, Hong Kong people are too busy even, too many choices, options. Entertainment and all kinds of … And you need to find out this group of people and then continue to engage them, sustain them, you know?
>
> *(Interview 2, NGO, Hong Kong, July 2011)*

Making Climateers an experiment therefore relied on a shared understanding of climate change as a problem of building a public constituency and sustaining engagement in reducing demand. While HSBC and WWF had previous experience of working together, Climateers required the reworking and reconfiguration of this alignment. For WWF, engaging corporate actors was central to their strategy of moving climate change up the policy agenda. Focusing on the way in which individuals were responding to climate change proved to be an area of common ground, and one in which a new reason to work in partnership could be found – the making of the Hong Kong carbon calculator:

> when we first designed the climate team as well as to engage Hong Kong people, we had to engage the corporates … they want to understand the [individual]. So [we] went out and start developing this brand new first Hong Kong local internet [carbon] calculator.
>
> *(Interview 2, NGO, Hong Kong, July 2011)*

The development of the carbon calculator served as a critical lynchpin in the making of Climateers, providing the basis for a new collaboration between HSBC and WWF on the one hand, and a new tool with which both could engage the public and demonstrate the potential of acting on climate change on the other. Reducing your carbon footprint became central to the organization of the website, the creation of a mobile app, the development of the notion of competition between Climateers, and a means through which both could demonstrate the value of achievements made through the programme. Subsequently, the Hong Kong government began to engage with the idea of carbon footprints, publishing their 2010 calculation that Hong Kong citizens emit 6 tonnes of GHGs per capita (a relatively low figure by the standards of global cities). In response, the calculator has been used to reframe the climate issue as one connected to broader patterns of consumption, reiterating the need for an active response to addressing energy consumption:

> we found that, on average, their [Climateers using the online carbon calculator] carbon footprint is 13.4 tonnes. … On average, it's quite high, yes. … [the Government] didn't count … the emissions from … aviation, but we count it too – so it is higher than the data provided by the government.
>
> *(Interview 3, NGO, Hong Kong, July 2011)*

In this case, the Climateers programme deliberately seeks to create an alternative calculation of what constitutes Hong Kong's carbon footprint through including emissions from aviation, and creating a different basis for imagining the problem that needs to be addressed in the city, providing the basis for the alignment of a particular coalition of actors.

Alongside the production of the carbon calculator as a technique which gave a common focus to the project, making Climateers an intervention that could work on the ground required the participation of the public. HSBC employees provided one source of volunteers. In addition, the Hong Kong Junior Police Corps and the Hong Kong Federation of Youth Groups provided a means through which the Climateers approach could be put to the test, at the same time developing some innovative forms of engagement. Having established its viability, the programme was then opened up to wider participation. The direct involvement of individuals in the programme required an approach that was engaging, in order that participation would be forthcoming. Two tactics proved central in this regard – *competition* and *immersion*. Competition features strongly across many of the initiatives undertaken in Hong Kong during 2007–2012 to engage individuals in programmes for behaviour change, such as the Friends of the Earth PowerSmart programme. In the case of the first phase of Climateers, competition focused on the development of proposals that would contribute to a low carbon lifestyle, with the promise of a reward for those individuals whose ideas were judged to be the most innovative:

> And then there is also a competition for them. ... they have a competition and then they can submit their own creative, innovative ideas on how to practice their low carbon lifestyle, in their daily lives. And then the winners can go to China for low carbon study tour.
>
> *(Interview 3, NGO, Hong Kong, July 2011)*

With over seventy submissions, the winning ideas included (HSBC 2013):

- wrapping the tap with a plastic band to reduce run rate of tap water;
- installing mirrors and reflective foil inside the lamp shade to add luminosity from a bulb with lower wattage; and
- cleaning cooking utensils with cuttlefish bone to reduce use of detergent and chemicals.

The intention was not that these ideas should be taken up or implemented through the Climateers programme, but rather the competition acts as a tactic through which the programme comes to engender participation and the generation of a debate about the alternative ways through which everyday actions could conceivably be conducted.

The idea of *immersion* as a means through which to engage the public with the idea and implications of climate change was also central to the way in which the programme became established. In its Climate Change Champions Scheme, developed as part of the HSBC Climate Change Partnership (and originally established in the UK with Earthwatch before being developed through its international business), HSBC engage their employees in a two week field course through which participants are given information about climate change and enrolled in collecting field data about the potential impacts of climate change on particular habitats. The use of the Mai Po nature reserve, a wetland habitat owned and managed by WWF, and the Ho Hai Wan Marine Park as the sites through which to conduct climate change education and training for the Climateers Ambassadors schemes echoes this approach, in which immersion in the environment considered to be at risk from climate change is seen as central to ensuring that the issue resonates, and that a form of connection is established to the everyday lives and actions of individuals. However, WWF did not see this as a tactic that was imported into Climateers from elsewhere, but rather was simply a matter of using their own resources to their advantage, given their position as manager of both sites. Wherever the idea of immersion initially stemmed from, it was regarded as critical to the Climateers programme, as a means of reconnecting to the environment and of ensuring that the programme was enjoyable:

> As well as [Ho Hai Wan], an ocean centre, to see the changes in the coral, because we have a [glass] bottom boat that allows us to do that. So you need to have fun, I mean, we are engaging the youth – rather than you are sitting in a classroom.
>
> *(Interview 2, NGO, Hong Kong, July 2011)*

While engaging members of the youth groups, with whom they had previously worked and which were already active in the climate change area, was regarded as fairly straightforward and as having gone as planned, the strength of response to the programme from the general public was unexpected, and largely attributed as being due to the opportunity to visit these conservation sites:

> *Respondent:* … what surprised us is the public. We received more than one thousand [applications for] enrolment in the public phase. … we mainly just promote [Climateers Ambassadors in] The Climateers e-news and the WWF e-news, and then this is already over one thousand enrolment in the public phase. So the … events really interest them the most because they have a chance to visit outlying villages …
>
> *Interviewer:* So would the seeing the coral reefs, that was the main …?
>
> *Respondent:* That was the main draw. Uh huh.
>
> *(Interview 3, NGO, Hong Kong July 2011, emphasis in original)*

Making Climateers an experiment therefore relied upon a particular problematization of climate change, the alignment of a set of social interests and specific urban conditions with this framing, and the development of tools and tactics that both sought to create the forms of knowledge and engagement required to address the issue through public engagement. In this case, however, the experiment was not stable, but rather required reinvention as the agenda developed and the social interests were reconfigured. In 2010, when the HSBC Climate Change Partnership came to an end, the alignment between HSBC and WWF shifted. At the same time, the Hong Kong government came to regard climate change as an issue that needs to be addressed in the city, and while most vocally advocating a response based on the development of low carbon sources of energy supply, also sought to promote energy demand reduction in the business sector and amongst the public:

> they [EPD] want to do something on climate change, then they prepared a budget to do public engagement, business engagement. So for business engagement they asked the four major chambers of commerce in Hong Kong to organise some activities for their members and then, for public

> engagement, we are one of the groups. Not many groups, you know, because of the tight schedule they propose and also some very difficult requirements for [it as well]. So we are, I think, only one of two, maybe, I guess, to do the public engagement work for them.
>
> *(Interview 2, NGO, Hong Kong, July 2011)*

Surprised by the level of interest in the Climateers Ambassadors programme, and seeking to innovate beyond the scope of the first phase of the experiment, WWF sought to refashion the programme as a means through which to foster the engagement of specific communities. This required realigning the scope and ambition of Climateers towards particular local communities, in Wan Chai and Sai Kung, and enrolling local organizations with a history of environmental concerns and of working with WWF – St James Settlement and TKO Environment Association. For St James Settlement, their involvement with climate change was sparked through the COME project (COME 2013) (a local currency initiative that seeks to engage local women in exchanging goods and services to support families on low incomes) and the Combat Climate Change Coalition. Their major focus is on addressing climate change as a matter of economic development: 'the major concept is: local production, local manufacturing and local consumption' (COME 2013: 272, translated). The TKO Environment Association also seeks to engage poorer, local, communities with environmental issues as a means of reducing the costs of living and providing opportunities to improve health and well-being. Engaging these social interests in Climateers required the tactic of competition to be reconceived as a means through which to develop new initiatives within the community. The resulting ideas are:

> quite different from the low carbon innovative ideas. ... Yes, quite different. Because they need to maybe organize an event for the community. Yes. So it's quite different. And then we give them a budget, so they have to work within the budget and then they can ... actually, they can submit the proposal in groups or individually.
>
> *(Interview 3, NGO, Hong Kong, July 2011)*

As well as retaining the immersion element of the Climateers Ambassadors programme that served to provide participants with an extra-ordinary experience and means of connecting to climate change, the second phase of Climateers also required that the issue was more explicitly embedded in everyday life, so that addressing climate change became ordinary and part of mundane activities, as it was perceived that 'Hong Kong people put more emphasis on convenience, speed' and, generally, 'things like [climate change] don't bother them' (Interview 2, NGO, Hong Kong, July 2011).

Maintaining social innovation through social networks

Ensuring that experiments work in practice requires not only the careful processes of problematization, alignment and assembly that go into making them, but also an active means through which they are maintained. We suggest that this process of maintenance is two-fold, requiring the often mundane, informal and improvised practices of *upkeep* on the one hand, and more strategic and purposeful tactics of *metabolic adjustment*, where experiments are used as a means through which to reconfigure existing forms of political, economic and ecological circulation (Chapter 2).

As a form of social innovation, maintaining Climateers as an experiment required upkeep of the social ties and everyday engagement of its participants. This was done primarily through the Climateers website, the design of which the WWF outsourced, at some expense, to ensure that it had the level of functionality and popular appeal they required. The end result 'looks, you know, very cool … compared to the EPD website … their website, people never go there' (Interview 2, NGO, Hong Kong, July 2011). The website was designed to appeal specifically to a younger generation, those initially targeted in the first phase of the programme, and to be a space through which Climateers could engage with one another independently of the partners:

> we have an interactive [eco-home] with low carbon office, you know, people can get some green ideas, low carbon ideas, from here. So we try to make it, all these kind of more interactive online web 2.0 kind of basis to outreach the younger generation.
>
> *(Interview 2, NGO, Hong Kong, July 2011)*

Designing such a website and providing relevant information and engaging material is of course not the same as ensuring that it is used in the way intended. The very qualities of a virtual space, as part of an increasing array of sites and social networks, meant that the viability of the website as a means of maintaining participation in Climateers was a concern, leading one partner to explain that they were considering 'viral marketing' as one way to increase the influence of the programme (Interview 3, NGO, Hong Kong, July 2011). The WWF also sought to engage with wider social networks, such as Facebook. Interestingly, while most of the main website (www.climateers.org) dates from 2010 and 2011, the Climateers Facebook page (www.facebook.com/climateers), which has over 1,300 'likes' at the time of writing, carries news and information through to August of 2012, suggesting that it was through this social network that Climateers was able to secure engagement beyond the Ambassadors programme.

As well as the use of virtual space, peer-to-peer networking through social networks was a key tactic designed to ensure the development and expansion of the Climateers programme. At the outset, members of the youth groups engaged as Climateers were seen to be critical to ensuring the spread and uptake of the programme:

> *Respondent:* [once] Facebook emerged and then you've got so many friends. And just then just one click and you can get the information to two thousand and three hundred people, all around, and then … inside ambassadors group, so they can invite their

friends to the Climateers, and they invite them to calculate their carbon footprint, just easy step – and then if their friends are interested they can look more at the website or they join our programme later on.

Interviewer: And have you got any sense of how many people have been referred to the programme that way?

Respondent: Over two thousand, yes.

(*Interview 3, NGO, Hong Kong, July 2011*)

Specifically, as part of their role, Climateer Ambassadors were asked to 'commit to introduce another pen friend to become Climateers' (Interview 2, NGO, Hong Kong, July 2011), a referral mechanism that was clearly central to the recruitment of so many individuals for the carbon calculator and online version of the programme. However, maintaining virtual spaces and networks requires real input in terms of time and resource. Translating the virtual presence of Climateers into the second phase of the programme was undertaken through the development of a mobile app to accompany the carbon calculator, funded as part of the support provided to the programme by the EPD, which three weeks after its launch in 2011 had attracted over 1,000 downloads. This reliance on the virtual world as a space within which to provide information, create networks and engage individuals in action was not readily translated into the reconfigured Climateers Ambassadors programme, given the different social and demographic makeup of the groups on the one hand, and their focus on the local community as a (real) space for action on the other. Nonetheless, social networks and peer-to-peer engagement remained central to the upkeep of the programme, enabling the circulation of ideas and examples beyond the constitution of the experiment. In the second phase, Climateers Ambassadors were also requested to recommend the programme to friends and family in order to develop its network:

> even without technology people just spread it. Nothing more. ... sometimes when we see them in different kinds of activities and then we will check with them and then see, and they say: 'yes, I told, I when I went to Yum Cha with my friend ...' ... from eating the meal.
>
> (*Interview 3, NGO, Hong Kong, July 2011*)

Alongside these processes of upkeep, what we term metabolic adjustment has also been central to maintaining the Climateers programme. Focusing on the need to change behaviour and reduce energy consumption has involved seeking to reconfigure the ways in which the costs and benefits of consumption decisions are made, and to introduce new technologies and practices, which in turn have the effect of changing the flows of resources required to sustain everyday practices. Two tactics have been deployed to this effect. First, the programme has sought to provide information and education about how to lead a low carbon life. One focus has been on informing Climateers about how their purchasing decisions may have a long-term impact in terms of energy and carbon. This was done through providing Low Carbon Living Tips and a Low Carbon Living Appliances Guide, which were informed by data on the use of electricity by different appliances in the city, provided by an expert third party. The Climateers programme and these assessments were positioned as a means of reconfiguring the flows of goods and resources. The applicances guide stressed not only the potential savings, but the relationship between making 'smart' decisions and putting into practice a collective response by taking leadership:

> Here you can compare and rank the appliances according to their energy efficiency, energy consumption, carbon emissions and learn more about the low carbon tips on the smart use of these appliances. By choosing appliances with low carbon impact and using them in a smart way, you are already part of a community of leaders that are tackling climate change!
>
> *(WWF Hong Kong 2013b)*

Alongside this rather conventional tactic of providing information and guidance, the Climateers programme explicitly sought to use forms of demonstration as a means of illustrating the capacity to live low carbon lives. Celebrities from different professions were enrolled as Climateers and used as a means of demonstrating what being part of the programme might involve, as this extract from the profile of Teresa Au, former Director of CSR for HSBC Hong Kong, illustrates:

> *Interviewer:* Do you think your footprint reflects your lifestyle?
>
> *Au:* In fact I have taken fewer business flights over the past year than usual! Unfortunately this coming year I may need to fly to quite a number of Asia-Pacific countries for regional meetings. But we already do regular teleconferencing and I will try to use it more often. At home the whole family consciously try to conserve energy and I take the public transport whenever possible.
>
> *Interviewer:* So you will work on that? What other Climateering actions are you going to do?
>
> *Au:* Raising that matter at the owners' corporation meeting will be on my to-do list! Come to think of it we do not have recycling facilities either, so I'll push for that and it should help us reduce waste. I am also going to work on energy efficient cooking methods, do more steaming, poaching and less frying. I am also going to be actively getting friends and colleagues to become Climateers!
>
> *(WWF Hong Kong 2013c)*

The tactic of demonstration was also critical to the engagement of local communities in the second phase of the Climateers programme. In Wan Chai, the winning proposal included a low carbon cooking competition in which Climateer

Ambassadors took part and had to select from a range of ingredients how to make a low carbon dish, which was then judged by a famous local chef.

> surprisingly they cook quite delicious meals and quite low carbon. Yes. And then the chef demonstrated two dishes that is low carbon: one is even without cooking, a salad, something like that. And then often they are quite surprised because they think that maybe low carbon is not delicious, something like that, but after all it's quite okay. And then we have to wash up for them because we want to teach them so low carbon tips. Yeah. And then the whole project is actually run by the winning participants. We are just assisting.
>
> *(Interview 3, NGO, Hong Kong, July 2011)*

There is apparently surprise in aligning good cooking with low carbon, and it is such surprise that maintains interest and facilitates the circulation of the experiment over time. In Sai Kung, the winning project was a cycling tour aimed at demonstrating the availability of cycling routes in the district and providing residents with information about options for renting and buying bicycles:

> We get involved cycling, like some sport we do together. ... This is somehow really let them get [to know] how to reduce those transportation expense, including emissions as well. That's an issue as well. So through this kind of impact we will actually involvement. Some of our residents never cycling before [until] through this type of workshop!
>
> *(Interview 4, Community based organization, Hong Kong, July 2011)*

Seeking to reconfigure flows of resource and economy around the notion of low carbon living through social innovation was addressed through the Climateers programme by deploying tactics of education and demonstration, notoriously fragile in achieving any such metabolic adjustment. Nonetheless, the presence of such intentions and tactics signals that, even for experiments focused on social innovation, the reworking of urban metabolic circulations is central, providing a critical objective, shared sense of purpose, and means through which to maintain the relevance and importance of the experiment. However, both in terms of upkeep and in seeking to reconfigure the use of resources, maintaining Climateers has been challenging – virtual networks have become disconnected, websites fallen into disuse, and community-based programmes completed with the end of specific funding arrangements. While the Climateers programme still has a presence within WWF and Hong Kong, its purchase in the city appears fragile.

Living experimentation: the practices of low carbon conduct

Essential to, and operating in tandem with, the processes of making and maintaining Climateers as an experiment have been the means through which it becomes lived. At the heart of Climateers and its rendering of the climate change problem as one connected to the over-consumption of energy in the built environment is the notion of the need to engender new forms of subjectivity in which living low carbon is normalized. In this instance, the entry point is the calculation of the carbon intensity of everyday living. Together, the carbon calculator, website, education and demonstration tactics employed through the programme seek to enable new forms of conduct in which individuals both govern themselves in relation to carbon consumption and seek to embed this practice in their daily lives:

> We want them to be practical in their daily lives. Even at home, at work too, they can … suggest to their supervisors at work, that maybe you can do this … We just want to get them well rounded. Not just thinking about 'me' as an individual – you are also at home, and then you are husband and wife, and then at work you are as an employee, but you can apply the system anywhere.
>
> *(Interview 3, NGO, Hong Kong, July 2011)*

The widespread appeal of the programme initially surprised the lead partners, but added impetus to their sense that there was a need to mobilize civil society in order to put climate change on the policy agenda within Hong Kong. Initial public engagement through Climateers reinforced the conviction that alternative forms of subjectivity in which being low carbon could be enacted and celebrated were possible. During the first phase of the programme, participants who were part of the Ambassadors programme reported a significant reduction in their carbon footprints, in the order of 20 per cent, further emphasizing the idea that being a low carbon citizen was possible. Although little evidence was gathered as to what participants had done and why, programme organizers felt that, while carbon and climate change may have initiated new forms of behaviour at home, their persistence was related to issues of cost and to the rediscovery of norms and practices about what energy use, particularly in relation to air conditioning, was like in the past. However, in its second phase the community-based Climateers Ambassadors programme found that, rather than leading to the adoption of new, low carbon forms of living, the programme struggled to adapt to daily lives in which consumption of carbon was already very low. To start with, the use of the carbon calculator assumed that people would have access to computers and the internet. It was also designed with relatively high carbon lives in mind:

> because they are more or less low income families here, so maybe the travelling part [by air or by ship] … is not so related to their life and not so applicable. For the electrical part, they don't really have that many different kinds of electrical appliances at their homes so it is also not so related.
>
> *(Interview 5, Community based organization, Hong Kong, August 2011)*

Rather than rejecting the calculation of carbon, and the programme of which it was a part, community leaders insisted that such forms of climate responsibility should be part of the environmental knowledge and education of their communities. This would, they suggested, enable citizens to take future decisions – over household purchases for example – with carbon in mind, and would contribute to the long-term education of the next generation about how and why environmental issues matter within daily life (Interview 5, Community based organization, Hong Kong, August 2011). Here, as elsewhere across the programme, the emphasis was not only on a form of low carbon subjectivity that is individualizing, where what matters is choice within particular circumscribed contexts (the office, the home), but also on the making of forms of low carbon society, within which communities are able to organize around low carbon, creating a constituency through which to advocate for wider political action on the issue and new forms of community-based governance:

> We want to extend to more communities, because they enquire, these people, why just Wan Chai and Sai Kung? So it depends on the funding and if there is funding we want to promote more to more communities ... we want the Climateers to act as a helping hand for us, to run different kinds of, maybe, mini programmes. ... Anywhere [in the city], maybe helping with something on the policy work too.
>
> *(Interview 3, NGO, Hong Kong, July 2011)*

FIGURE 9.2 Air-conditioning is integrated into the urban fabric and everyday practice of Hong Kong. Photo © H. Bulkeley.

Despite the promise it seems to hold, seeking to create new forms of low carbon subjectivity is not without challenge or contestation. Through the process of peer marketing, Ambassadors experienced direct resistance to their requests for friends and family to become involved by making changes to their everyday lives, such as reducing the amount of meat they were eating, using less electricity or travelling less often:

> [they say] 'meat is delicious' or 'I just feel comfortable in air conditioning, why I need to stop?' … But we always encourage them to say, okay, you can say to your friends: when you use energy efficient appliances … or something like that, you can help them to save money.
>
> *(Interview 3, NGO, Hong Kong, July 2011)*

Such explicit forms of resistance to being a low carbon citizen were also accompanied by the prevailing norms of energy consumption in Hong Kong. In the context of monopoly energy provision and relatively low prices for electricity, an urban environment which has been developed around the extensive use of lighting and air conditioning, and cultural norms concerning everything from dress codes to the use of indoor retail spaces as a form of leisure, carbon consumption is deeply embedded within the city (Figure 9.2). From 'the first day that you are born … you will be kept in an air conditioned room' (Interview 6, NGO, Hong Kong, July 2011); in large parts of central areas of the city people 'don't even need to step in the street. I have a whole map in my mind … I know the connections for the buildings. All of this is air conditioned' (Interview 6, NGO, Hong Kong, July 2011). As stakeholders recognize all too well, gaining purchase on the carbon lives lived through Hong Kong requires a reconfiguration of the conditions through which they are lived. Creating new forms of low carbon subjectivity through programmes such as Climateers may be one means through which this is enacted, offering an alternative sense of both the self and of the city, but it remains marginal to mainstream forms of urban development and everyday life.

Impacts and implications

The Climateers programme formed one node in a web of interventions developed in Hong Kong by a loose coalition of businesses and NGOs that sought to problematize climate change as a matter of energy demand and to create constituencies through which to mobilize the policy agenda. As a form of social innovation, it signifies what Paterson and Stripple (2010: 341) have termed the emergence of a 'my space' form of climate governance, in which forms of carbon calculation are used as the basis for establishing new forms of self-governance. Such interventions and experiments may be important in their own terms. In this case, the Climateers programme attracted widespread membership, recruiting over 15,000 people who signed up to be a Climateer, over 5,800 who used the carbon calculator, and over 3,000 downloads of the carbon calculator app. For the 180 people who became

Ambassadors in the first phase of the programme, self-reported figures suggest they reduced their carbon footprint by over 20 per cent. In the local communities that participated in the second phase of the programme, most of whom lived relatively low carbon lives, the programme served to sustain access to environmental education and provide a further means through which community organization around issues of local economy, sustainable transport and environmental consciousness could take place. As an urban climate change experiment, Climateers succeeded in aligning a diverse set of interests and elements into a coherent space for intervention, creating the means through which to enrol diverse publics, and enabling it to demonstrate that forms of low carbon living in Hong Kong were possible. This success was sustained through the creation of low carbon publics, in the form of the virtual network of Climateers, the social network of Ambassadors, and the local networks of community-based organizations. For the virtual and social network of climate publics, Climateers provided the catalyst for the formation of these communities, the traces of which can still be seen – the Facebook page continues to attract comment and support, for example – in the case of the local networks, it served to reinforce existing ties of community organization.

The work of experimentation also travelled beyond the boundaries of the intervention itself. In Hong Kong, carbon calculation has developed apace since being used in the Climateers programme. The first item listed by the Environmental Protection Department as what 'you' can do to address climate change is calculating your carbon footprint, and the site now lists 10 different carbon calculators. In keeping with the government's focus on the need to enhance energy and carbon auditing, over 30 of the 300 projects funded by the EPD's Environmental Conservation Fund explicitly state that they include some form of audit or footprint process. Carbon calculation has, through experiments such as Climateers and a growing focus on the voluntary regulation by the private sector of the built environment within government (Chu and Schroeder 2010), become central to the means through which the governing of climate change is taking place in the city. In this respect, Climateers also succeeded in its ambition to normalize the issue of energy demand reduction as part of the climate change problem, and to demonstrate particular techniques that could be applied to such a problem context.

The Climateers programme and similar initiatives also served as an important means through which to sustain the coalition of private actors who were mobilizing around climate change. The late 2000s witnessed the rapid proliferation of institutions, such as the Climate Change Business Forum, reports, events and initiatives undertaken by a range of private actors in the city and intended to create policy momentum. Within this melee, interventions such as Climateers that were able to demonstrate concrete results were certainly valued. At the same time, interviewees commented on the relatively ephemeral nature of the initiative, the extent to which it was able to have purchase on the everyday lives of Hong Kong citizens who were seemingly wedded to an air-conditioned, high-energy culture, and how it served not as a means to persuade the government to take action but rather as a means through which such responsibilities might be avoided. Such views speak to the fact that the stated goal of Climateers to engender an urban response to climate change in Hong Kong has not yet been achieved. Almost three years after the consultation document was published, Hong Kong's position on responding to climate change remains ambiguous. Individual programmes and strategies are in place, many of which have been criticized for their lack of ambition (Chu and Schroeder 2010; Francesch-Huidobro 2012), but the government remains uncommitted to a specific Hong Kong position. In this context, it appears as if the ambition of the loose coalition of businesses and NGOs in Hong Kong advocating for further commitment has been unsuccessful. Yet, at the same time, climate change remains on the policy agenda, with the matter of attending to energy demand receiving support in the form of funding and governmental programmes. The situation is therefore one of limbo, a context in which experiments have flourished and gained momentum, but where their longevity is constrained by the continual making and remaking of the policy agenda and the ongoing normalization of high levels of energy consumption as underpinning the economic development and success of the city.

References

Chu, S.Y. and Schroeder, H. (2010) 'Private governance of climate change in Hong Kong: an analysis of drivers and barriers to corporate action'. *Asian Studies Review*, 34(3): 287–308.

COME (2013) *Community Oriented Mutual Economy*. Available online at http://wikibin.org/index2.php?option=com_content&do_pdf=1&id=33196 (accessed December 2013).

Francesch-Huidobro, M. (2012) 'Institutional deficit and lack of legitimacy: the challenges of climate change governance in Hong Kong'. *Environmental Politics*, 21(5): 791–810.

Gouldson, A., Hills, P. and Welford, R. (2008) 'Ecological modernisation and policy learning in Hong Kong'. *Geoforum*, 39(1): 319–30.

HKSAR (2010) *Hong Kong's Climate Change Strategy and Action Agenda: Consultation Document*. Environment Bureau, Hong Kong Special Administrative Region, September 2010. Available online at http://www.epd.gov.hk/epd/english/climate_change/files/Climate_Change_Booklet_E.pdf (accessed 10 February 2014).

Hoffmann, M. J. (2011) *Climate Governance at the Crossroads: Experimenting with a Global Response after Kyoto*. Oxford: Oxford University Press.

HSBC (2013) *Support Climateers Programme on Low-carbon Living*. Available online at http://www.hsbc.com.hk/1/2/cr/community/projects/climateers (accessed 10 February 2014).

Paterson, M. and Stripple, J. (2010) 'My Space: governing individuals' carbon emissions'. *Environment and Planning D: Society and Space*, 28(2): 341–62.

The Climate Group (2010) *A Low Carbon Vision for Hong Kong*. Hong Kong: The Climate Group. Available online at http://www.theclimategroup.org/what-we-do/publications/a-low-carbon-vision-for-hong-kong-discussion-paper/ (accessed 10 February 2014).

——(2013) *HSBC Climate Partnership*. Hong Kong: The Climate Group. Available online at http://www.theclimategroup.org/programs/hsbc-climate-partnership/ (accessed 10/February 2014).

WWF Hong Kong (2013a) *About Climateers*. Available online at http://www.climateers.org/eng/contents/about_climateers.php (accessed 10 February 2014).

——(2013b) *Low Carbon Living Appliances Guide.* Available online at http://www.wwf.org.hk/en/whatwedo/footprint/climate/examples_of_individual_actions/climateers/beaclimateer/appliancesguide/ (accessed December 2013).

——(2013c) *Teresa Au.* Available online at www.wwf.org.hk/en/whatwedo/footprint/climate/examples_of_individual_actions/climateers/beaclimateer/whos_climateering/teresa_au/ (accessed 10 February 2014).

10

CREATING A LOW CARBON ZONE IN BRIXTON, LONDON, UK

with Sara Fuller

Introduction

The London Borough of Lambeth is one of the most densely populated of the 14 local authorities which make up inner London, with a population of around 300,000 (London Borough of Lambeth 2013a), and it is comprised of multiple urban centres including Brixton. Known for its vibrant, multicultural and diverse communities (London Borough of Lambeth 2013b), over the past five years multiple interventions to respond to climate change have been conducted in Brixton. In this chapter, we examine how the Mayor of London's Low Carbon Zone (LCZ) was deployed in Brixton, showing how this has established new networks and initiatives through which the low carbon transition is being orchestrated in London. The LCZ has facilitated a shift from instrumental, top-down interventions to foster sustainable energy practices to an ongoing process of organic self-regulation that built upon previous experiences of sustainable food management. This shift has challenged a priori assumptions about who could intervene and how in achieving a sustainable energy system for Brixton.

In the first part of the chapter, we trace the development of climate change policy responses in London, highlighting the key role of the Greater London Authority (GLA) in articulating climate change as an urban agenda. Next we turn to the specific intervention of the Low Carbon Zones, and the ways in which they reflect a growing interest across the UK in area- and community-based approaches to mitigation. In the third part of the chapter, we consider the case of Brixton Low Carbon Zone in detail, examining how it was made, maintained and lived. Making the Brixton LCZ involved building links between global climate change mitigation objectives and local concerns about energy poverty and community cohesion, in such a ways as to support an integrated agenda focused on energy and resource maximization. Processes of maintenance encouraged the growth of community involvement activities, which also led to a change of focus from individual behaviour change to collective action and social enterprise as the LCZ was lived, creating new low carbon subjects in the process.

Establishing climate change as an urban agenda in London

Historically, the UK has been at the forefront of the international response to climate change. Since the late 1980s, when then Prime Minister Margaret Thatcher acknowledged in a speech to the Royal Society that 'it is possible that … we have unwittingly begun a massive experiment with the system of this planet itself' (Thatcher 1988), the UK supported the development of climate change science, through the activities of the Met Office and Sir John Houghton's role as Chair of the first IPCC Working Group I, and international diplomacy for the development of the UNFCCC and Kyoto Protocol. Domestically, so-called 'dash for gas' in the 1990s served to decarbonize the UK's energy system such that, with Germany, it was one of the only countries to be on track to meet the targets agreed by the European Union under the 1997 Kyoto Protocol. Subsequently, a number of policy innovations, including a fuel-duty escalator, and programmes to increase the energy efficiency of buildings and encourage renewable energy were adopted. However, by the early 2000s there was concern that the UK was no longer on track to meet its ambitious target of reducing greenhouse gas (GHG) emissions by 20 per cent below 1990 levels by 2010, and the Royal Commission on Environment and Pollution called for a long-term goal of reducing emissions by 60 per cent below 1990 levels by 2050. In 2006 the UK Climate Change Programme formally adopted this goal, and in 2008 the Climate Change Act bound the UK government to pursue it under the oversight of a Climate Change Committee and through five-year carbon budgets.

If national-scale policy development has driven action on climate change in the UK, the role of municipal authorities has been somewhat muted. While several local authorities were amongst those pioneering a local response to climate change in the early 1990s, with voluntary targets for reductions in energy use and GHG emission reductions from vehicle fleets and buildings (Bulkeley and Betsill 2003; Bulkeley and Kern 2006), for the most part local authorities were seen to lack the capacity and resources to respond effectively to the issue. Momentum for such voluntary approaches grew in the early 2000s, through the development of the Nottingham Declaration and the network of local authorities which signed up to its commitments, but for the most part climate change was considered a relatively minor issue on municipal agendas, even though many of the 32 Borough Councils in London had been quite progressive. This agenda was transformed by the election of Mayor Livingstone and creation of the GLA, such that climate change came to be regarded as a strategic issue for the economic development of the city and for its standing on the global stage.

The successful introduction of the London Congestion Charge in 2003 had demonstrated that there was broad support for measures with both economic and environmental benefits, and London's 2004 Energy Strategy articulated its intention to adopt a long-term approach to significant emission reduction targets, in line with the report of the Royal Society, such that 'London should reduce its emissions of carbon dioxide by 20 per cent, relative to the 1990 level, by 2010 as the crucial first step on a long-term path to a 60 per cent reduction from the 2000 level by 2050' (GLA 2004: x). The Energy Strategy focused on the use of the planning powers vested in the GLA to promote the use of micro-generation in new developments, and the formation of partnerships to assess the potential for enhancing energy efficiency in the city (Bulkeley and Schroeder 2008). Following the re-election of Mayor Livingstone in 2004, a more ambitious approach was adopted including the formation in 2005 of the London Climate Change Agency as a vehicle through which to develop new energy technologies in the city, and the development of the 2007 Climate Change Strategy with the ultimate goal of 'stabilising CO_2 emissions in 2025 at 60 per cent below 1990 levels, with steady progress towards this over the next 20 years' (GLA 2007: 19). Despite recognising that much of the capacity to realize such targets lay in the hands of the national government, Livingstone sought to pursue an ambitious and long-term approach to addressing climate change in London, primarily through decentralization of energy production and an increased level of self-sufficiency in the production of energy resources in the city, as well as various schemes that sought to enrol businesses and households in energy efficiency programmes.

At the same time as seeking to develop an ambitious approach to climate change within London, Mayor Livingstone and his Deputy Nicky Gavron worked with The Climate Group, a London-based transnational organization seeking to promote action on climate change, to establish what would become known as the C40 Cities Climate Leadership Group (C40). C40 started as a side event to the G8 Meeting on Climate Change in 2005, bringing together Mayors from the global cities located in the G8 countries, and rapidly expanded in partnership with the Clinton Climate Initiative. When the Conservative Party came to power in 2008 with the election of Mayor Boris Johnson, there was concern from some quarters that the new administration would have little time for climate change action. However, in fact it has remained high on the political agenda, in part because Mayor Livingstone had already positioned London as a leading city in the global response to climate change, attracting support from the city's economic elites who were concerned with carbon finance and insurance against the impacts of climate change. In addition to the continued focus on developing decentralized energy technologies, programmes for energy efficiency and community involvement have become focused around the Re:New programme for domestic energy retrofit, the Re:fit programme for public buildings, and the continuation of schemes for businesses to increase their energy efficiency.

Of particular interest in the development of London's approach to tackling climate change has been a new focus on the role of communities in responding to the issue, mirroring the greater mobilization of community- and area-based responses to climate change across the UK (Mulugetta *et al.* 2010; Walker 2011).

In the rest of this chapter, we examine one such case – the Brixton Low Carbon Zone – to consider the multiple forms of interest and agency that are enrolled in area-based climate change experiments, and analyse the ways in which it has evolved from a programme established 'from the top down' by the Mayor's office, to become a project that is embedded in different forms of climate change response within this area of London.

Low Carbon Zones and the urban politics of climate change

Initially announced in July 2008 shortly after his election, the London Low Carbon Zones Programme was formally launched in May 2009 by Mayor Boris Johnson: 'These energy busting zones will create an armada of flagships across London, focused on finding the most effective ways to rapidly cut carbon and slash energy bills' (GLA 2009). The Mayor pledged £3 million of funding to support up to ten Low Carbon Zones across the city (see Box 10.1). The primary objective of the Low Carbon Zones Programme was to deliver 'rapid carbon savings from buildings in the zones and the development of models that drive long-term carbon savings', and in addition to contribute to the 'mitigation of fuel poverty, promotion of sustainable lifestyles and lower carbon footprints and regional skills development and other positive social outcomes' (GLA 2009: 16). The intention of the scheme was to support a community-based approach for testing innovative measures for bringing about a 20.12 per cent reduction in carbon emissions before the London 2012 Olympics and to put in place mechanisms to achieve a further 60 per cent reduction by 2025.

BOX 10.1 LOW CARBON ZONES

A Low Carbon Zone (LCZ) is a local geographic area with concentrated activity to try out innovative ways of reducing CO_2 emissions. The LCZ programme will bring together local authorities, business and industry, public sector and community organisations to develop, demonstrate and measure the impact of initiatives that create a big reduction in CO_2 emissions.

The initiatives comprise technological innovations and programmes accelerating established measures and encouraging behavioural change, covering areas such as:

- Residential and commercial building retrofit solutions including 'easy measures' such as loft and cavity wall insulation as well as 'harder measures' such as solid wall insulation and double-glazing and measures that are still under development like smart meters
- Decentralised energy to produce heat and electricity locally for example using energy from waste

- Use of renewable energy and microgeneration
- Community engagement, events and marketing to drive behavioural change including transport mode shifts to cycling and walking

LCZs act as exemplars for London, showcasing technological solutions and catalysing long-term reductions in CO_2 more broadly across the capital.

Concentrating climate change activity at the neighbourhood level was thought to provide a number of advantages over other approaches, particularly in terms of realising additional benefits locally, achieving the behavioural change regarded as necessary to meet London's goals, and creating more efficient responses than those focused on individuals or households:

> It offers communities a real freedom to tailor-make plans to fit with the specific challenges and opportunities of their neighbourhood. Giving communities the chance to design and manage their plans to reduce carbon emissions means they can better integrate services and deliver more meaningful behaviour change in their neighbourhood. A local approach also provides economies of scale in terms of raising funds and speeding up delivery by partners.
>
> *(GLA n.d. b)*

Local authorities were invited to bid for funding for an area of up to 1,000 residential, commercial and public sector buildings, and in September 2009 ten Low Carbon Zones were announced in the London Boroughs of Haringey, Islington, Westminster, Barking & Dagenham, Richmond, Merton, Sutton, Lambeth, Southwark and Lewisham (see Figure 10.1). In total, the Zones covered over 13,000 residential properties, around 1,000 shops and businesses, and 20 schools, as well as hospitals, places of worship and community centres (GLA 2011: 135). Within Lambeth, Brixton was selected as the LCZ for a number of reasons:

> Brixton has the most suitable range of buildings and existing projects. Brixton also has high levels of deprivation, fuel poverty and a significant need for carbon reduction measures. In addition, Brixton has the potential to attract funding and there is scope to build on existing community engagement initiatives.
>
> *(London Borough of Lambeth 2010)*

Confirming the Borough's decision, one respondent explained that Brixton was appropriate in terms of 'fitting all of the criteria' and 'being able to address a lot of the issues' (Interview 1, Municipal Authority, London, April 2011).

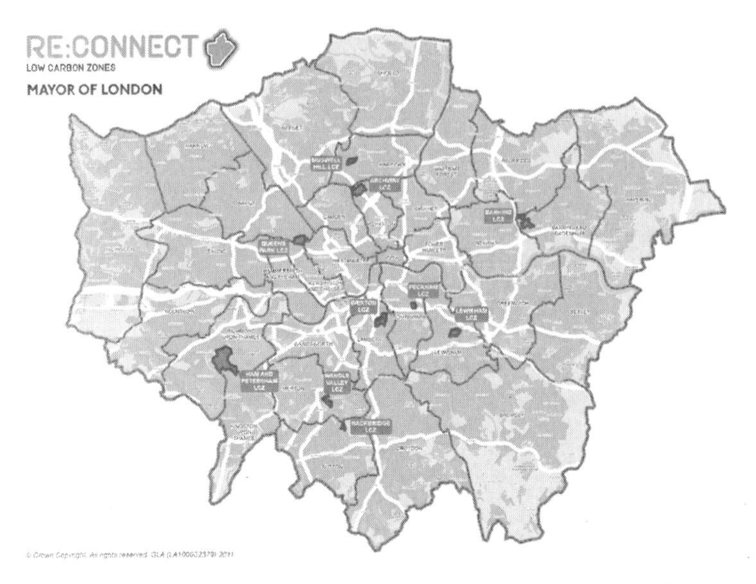

FIGURE 10.1 The ten London Low Carbon Zones. Drawing © GLA, used with permission.

The Brixton LCZ contained around 720 buildings (approximately 3,500 properties) including ten high-rise and 36 low-rise blocks, street properties (social and private housing), and commercial and public sector buildings (Figure 10.1). In Brixton, as with each LCZ, the Zone was led by the local authority with a range of partners (local authorities, private sector partners and community organizations) and the intention of leveraging private sector investment from energy suppliers. In recognition of this, the programme was rebranded as RE:CONNECT, drawing attention to the importance placed on linking the Mayor's office to the communities, local authorities and private sector interests engaged in the scheme. This notion of connecting communities, local authorities and the private sector was critical to assembling the LCZ in Brixton, the means through which it sought to entrain climate change with wider social and environmental concerns in the neighbourhood, and the processes through which it came to be lived and practiced locally, and through which the experiment was made, maintained and lived.

Making Brixton a Low Carbon Zone

The making of climate change experiments requires processes of problematization and alignment through which different actors and entities come to share an understanding of the need and scope of intervention and configure their interest and agency in relation to one another. However, as we argued in Chapter 2, such processes by themselves are not sufficient for animating climate change experiments, and alongside them we suggest that both forms of calculation and of persuasion are required, such that the experiment becomes regarded as visible, practicable and effective, on the one hand, and as worthy of support and investment on the other.

Three processes were of particular importance for the Brixton LCZ: the convergence of local and citywide policy discourses around climate change; the configuration of the LCZ as something which required the input from a wide constellation of actors; and the process of bidding for and creating the LCZ itself, which provided a means through which the immanent alignment around climate change could be made practicable and which gave those involved in addressing climate change in the area a reason to engage with one another.

As discussed above, climate change rose rapidly on the agenda in London during the 2000s bringing high visibility to the issue amongst policy makers and the public. Within this context, the development of the Brixton Masterplan provided a means through which climate change came to be related to the city:

> The Masterplan focuses on the long term sustainability for Brixton – environmental, economic and cultural. We want to establish Brixton as a world leader on sustainability issues such as zero-carbon and zero-waste development. In doing so, the plan adopts the 'One Planet Living' principles – an ambitious set of guidelines for sustainable development developed by the BioRegional and the WWF – as a core set of aspirations and structuring devices.
>
> *(London Borough of Lambeth 2009a: 7)*

These aspirations were clearly reflected in Lambeth's application for a LCZ, which argued that 'Brixton is a vibrant hub of carbon reduction activity' (London Borough of Lambeth 2009b: 1). The application contended that being 'low carbon' was already embedded in the local authority, and the LCZ could address multiple agendas, such as the Affordable Warmth Strategy, Housing Strategy, Local Area Agreement, Sustainable Community Strategy, Economic Development Strategy, Local Development Framework and Community Safety Strategy (London Borough of Lambeth 2009b). As one interviewee explained:

> While putting in the bid we had a lot of internal partners ... the housing team, who has a remit around addressing a lot of the fuel poverty issues and retrofit projects, CESP funding ... work with the region and enterprise team ... with the waste teams ... and community engagement officers – they're based in the sustainability unit. It was a very cross-cutting project. So while putting in the bid we had to bring all the teams and all the possible aspects of it together.
>
> *(Interview 1, Municipal Authority, London, April 2011)*

Yet, at the same time as staking a claim to be leading in the area of carbon reduction, the bid also recognized the need to develop a specific focus for activity in order to pursue the wider ambitions of the Borough: 'This area has been chosen because without intensive carbon reduction activity Brixton cannot achieve its goal of becoming a green hub. The purpose of the Brixton LCZ is to address these challenges and secure Brixton's future as a leader in carbon reduction' (London Borough of Lambeth 2009b: 2). In this manner, the LCZ was envisaged as something of a crucible, a space within which multiple agendas could be melded together and intensified as a means of giving traction to the broader sense that climate change was an urban problem to which some form of response was required. In this sense, practices of calculation were required to link broader objectives to the administrative area. In addition to the convergence in policy agendas achieved between the Mayor of London and the Borough of Lambeth, and within the council itself, the second critical process that was involved in the making of the LCZ as an experiment was the alignment of multiple actors beyond the local authority. The brief provided by the Mayor was clear that LCZs should be led by local authorities but should also connect to private sector actors (particularly energy providers) and the community at large. The Brixton LCZ reflected these intentions to the letter, such that these actors and their alignment is central to any description of the project:

> Brixton Low Carbon Zone is a complex multi-agency and multi-sector project. Social housing providers such as Lambeth Living, United Residents Housing and the Metropolitan Housing Trust are involved in sustainable energy installations and resident engagement ... Organisations such as Transition Town Brixton, Remade in Brixton and Invisible Food are also involved in engaging communities within the Zone. Negotiations are also ongoing with energy companies regarding Community Energy Saving Programme monies for the Loughborough Estate.
>
> *(London Borough of Lambeth 2010)*

The LCZ also involved the alignment of global and local actors, two of particular significance. First, Transition Town Brixton (TTB), a group which is part of the wider Transition movement and which aims to 'support community-led responses to peak oil and climate change, building resilience and happiness' (Hopkins and Lipman 2009: 7). In line with the wider Transition movement, the emphasis in TTB is 're-localising' and building local community resilience. As a result, the projects undertaken under the umbrella of TTB are broad and include, for example, the Brixton Pound (an alternative local currency which was the UK's first local currency in an urban area) and multiple projects aimed at growing food locally. Interviewees noted the importance of the visionary nature of the transition process, which focuses on possible solutions rather than just viewing climate change as a 'problem' to be addressed. This 'positive problematization' of climate change and the longstanding relationship between TTB and the Borough Council was a key part of Lambeth's bid for the LCZ, since TTB played a key role in enabling the LCZ to develop into a project around which multiple different agencies could gather. A second key actor, United Residents Housing (URH), was – prior to council closure in 2013 – an arms-length management organization owned by Lambeth Council with responsibility for four Tenant Management Organisations within Lambeth.

One of these, the Loughborough Estate, is within the boundaries of the LCZ and so has been central to much of the work undertaken within it. The Loughborough

Estate has a mix of properties including large ten-storey tower blocks, London County Council red brick, walk-up blocks, maisonettes and terraced houses. These required significant retrofits to reduce energy use and associated GHG emissions, providing a rationale for investment by the LCZ, while in turn by engaging with the LCZ, URH was able to successfully attract investment for capital improvements through the Community Energy Savings Programme (CESP) on the Loughborough Estate. CESP was an energy saving programme which ran from 2009 to 2012, funded by an obligation on energy suppliers and on electricity generators. It aimed to reduce fuel bills of low-income households across the UK and improve the energy efficiency of the existing housing stock in order to reduce the UK's CO_2 emissions through three key principles: partnership working; a whole house, intensive approach; and low-income targeting (DECC 2011). Such a programme facilitates concrete interventions to mobilize investment within the local context. This shows the mutuality involved in aligning different socio-technical entities in the urban context in relation to climate change experiments, and how, in the process of making experiments, they become already embedded in wider circulations of policy, economy and investment.

A third critical set of processes included those of calculation and persuasion involved in the design and implementation of the scheme itself. As discussed above, the overall aim of the LCZ programme was to test innovative measures for reducing carbon emissions. Reflecting the convergence of agendas in Lambeth between economic development, poverty alleviation and climate change, and the alignment of critical actors and socio-technical urban networks through which the LCZ came to be regarded as a potential intervention, retrofit activity was identified as the key strategy to reduce emissions in the Brixton LCZ. The Loughborough Estate in particular was one of the least energy efficient estates in Lambeth and the LCZ application set the goal of transforming the estate into an example of best practice in sustainable energy. At the time the bid was submitted the expectation was that the funding for capital programmes needed to secure the carbon savings would be levered in from private sector partners through the CESP. Proposed CESP measures included roof renewals, solid wall insulation and solar PV units (London Borough of Lambeth 2009b), funded by £2.8 million of investment from E.on Energy plus £3 million from URH. Because projections suggested that private sector funding could deliver the retrofit programme, the £300,000 budget from the GLA was able to be allocated to other processes. In particular, the bid emphasized community development:

> To maximise the benefits of installing carbon reduction measures, the Brixton LCZ will focus on behaviour change and building social capital to simultaneously tackle fuel poverty and social exclusion. Residents will be engaged through initiatives that address sustainable energy use, food growing and recycling in ways that are enjoyable, practical and bring the community together.
>
> *(London Borough of Lambeth 2009b: 2)*

To enable this behaviour change activity and community capacity building, the Brixton LCZ incorporated higher levels of staffing than other LCZs, employing a project officer, waste officer and community engagement officer. This was seen as 'added-value', as one interviewee noted:

> I feel they were fairly confident at the time that they were putting in the bid that all of these other programmes would take place, like CESP would take place, various other inputs … in terms of boiler replacements, and some roof insulation work. So in terms of residential properties they were fairly confident that … through Lambeth's capital allocation programme, there would be work happening within the zone.
>
> *(Interview 1, Municipal Authority, London, April 2011)*

In this way, climate change action in Brixton was driven by the desire to develop community cohesion and behaviour change, and was effective because of investment and capital expenditure in retrofitting the existing building stock. Bringing climate change and community cohesion objectives into alignment was key to making the experiment compelling to potential partners in the delivery of LCZ, since by problematizing climate change as a cross-cutting issue, the goal of reducing carbon emissions through retrofitting was made amenable to investment. Paired with the way the experiment produced social capital in the community, this served to effectively embed the LCZ within Brixton.

Maintaining the Brixton Low Carbon Zone

As with the other climate change experiments discussed in this book, the Brixton LCZ required ongoing processes of repair in a social, technical and political sense. At the same time, the presence of the LCZ reworked wider circulations across the Borough and the city in three areas that were required to maintain it: its visibility and tangibility in diverse contexts; its integration into the community; and its patterns of responsibility allocation.

First, the design of the LCZ included specific interventions which served to maintain and reproduce the Zone as a visible area of experimentation in the city. The most extensive form of intervention framing the LCZ was the retrofitting of the Loughborough Estate, funded through CESP. Lambeth Council committed £7.5 million as match funding to draw down £2.5 million of funding from the energy supplier E.on. This was used for a large-scale energy efficiency retrofit programme for 845 homes on the Loughborough estate (started in January 2012) which included: insulated flat roof system to tower blocks; pitched roof repairs and dormer renewals to low rise blocks; window replacements; insulated cladding to external walls; new gas central heating; new front entrance doors and decorations in external and communal areas. In seeking to maintain this process of retrofitting as one which was centrally connected to the reduction of carbon in the Borough, as well as meeting long term social and economic needs, previously hidden interventions were made more explicitly visible and new forms of responsibility were negotiated, such that URH came to see themselves as having a role to play in low carbon transitions:

> Certainly from our point of view it's looking at what we can do because after all we do major works, we do a lot of stuff so we have an obligation to try and be responsible and to address wherever we can climate change issues … so for example in CESP … we're changing G rated boilers for A rated boilers … so that's all the stack of carbon work and energy efficiency stuff so … it's up to us to actually do it.
>
> *(Interview 7, Municipal Authority, London, May 2012)*

> Interestingly possibly stimulated by the Low Carbon Zone, URH may have upped its game … they obviously have to develop their estate, their buildings, they have various government things they have to do like the decent homes which is the minimum standards that they have to achieve and it involves quite a bit of capital investment but I think as a result of the low carbon zone they've probably been looking at what is the bang they get for their buck … we were talking to URH about modifications before the low carbon zone came along.
>
> *(Interview 5, CBO, London, November 2011)*

Second, the LCZ placed considerable emphasis on community integration. Alongside the capital investment programme, as discussed above, the Brixton LCZ highlighted community-led responses, particularly those concerned with achieving behaviour change and building social capital in relation to energy and low carbon issues (Table 10.1). To achieve this the LCZ was also aligned with pre-existing projects – such as Green Doctors, the Green Community Champions and the Friends of Loughborough Park – that were enrolled into and partly managed under the auspices of the LCZ (Figure 10.2). In relation to reducing carbon footprints and testing new approaches to behaviour change, two other projects also became significant: the Community Draughtbusters and the Brixton Energy project. Participants saw this form of locally-embedded experimentation as one of the key benefits of the funding:

> I must say GLA were really good funders in the way they manage it. They have acknowledged the fact that each zone is an individual area with its own uniqueness and each of the zones has been using their funding in a very different way. I know some of the other zones are using the funding to match fund from other larger capital programmes … whereas the way we've used our funding is largely around being able to employ, say myself, the community engagement officers – so more people based.
>
> *(Interview 1, Municipal Authority, London, April 2011)*

> Last year I knew we would have a bit of an underspend … I was able to pool it and put a pot together – £7,000 to continue the Green Doctor project, and provide a grant to our community draughtbusters … So they are very flexible, [as] that wasn't in the plan.
>
> *(Interview 1, Municipal Authority, London, April 2011)*

TABLE 10.1 Examples of key projects within the Brixton Low Carbon Zone

Project	*Overview*
Green Community Champions	A facilitation, support and networking service for green community groups and individuals launched in 2009. While not funded directly through the LCZ, one of the LCZ's key objectives was to facilitate the Green Community Champions network as a mechanism of increasing community capacity to deliver projects as well as increase community cohesion.
Friends of Loughborough Park	The LCZ has supported local projects emerging from community strengths such as food growing and community gardening. One of these groups is the Friends of Loughborough Park who have created raised vegetable beds in Loughborough Park and hold growing sessions each week. The group have funding which they planned to spend on a shed, tools and fruit trees. While the initiative is not directly part of the LCZ it links to wider agendas and one of the founders is a Green Community Champion.
Green Doctors	A 6-month scheme funded by Groundwork via the Future Jobs Fund which provided a placement for five people to work as Green Doctors. It pre-dated the LCZ funding but it was supported and managed in practice by the LCZ community engagement officer. The Green Doctors completed 150 home energy visits, providing energy advice and the installation of energy saving measures such as energy saving light bulbs, radiator panels and an electricity monitor. In addition, they delivered draught proofing to over half of those homes visited, which led directly into a series of curtain lining workshops. Three of the five Green Doctors recruited secured full time employment as a direct result of the scheme. The Green Doctors scheme was continued via the Community Draughtbusters.
Community Draughtbusters	Originating from a local Climate Action Network, there have been a range of activities under the remit of community draughtbusters including advice on saving energy and money, installation of energy saving light bulbs, radiator panels and energy monitors, draughtproofing of doors and windows and a series of curtain lining workshops. Community Draughtbusters was initially funded with a grant from the Brixton Low Carbon Zone with one of the key objectives being to develop draughtbusters into a standalone social enterprise.
Brixton Energy	A community owned energy project developed as a grassroots project under the auspices of the LCZ and building on ongoing work by TTB's Building and Energy Group. The vision of the scheme is to create a greener future for Brixton by: generating energy; increasing energy resilience and security; raising awareness about energy efficiency and tackling fuel poverty; and providing training and employment for local people. Brixton Energy operates as a mutual society and gathers investment through the sale of shares to community members. Since the first share offer in 2012, solar panels have been installed on a second building, with a third share offer in progress. Brixton Energy has been recognised nationally and internationally for the pioneering nature of the project.

The provision of funding with relative flexibility provided the potential to develop appropriate responses to the broader challenge of climate change that, while being suited to the local context, were tangible and visible. Yet, in the context of challenging carbon reduction targets, the flexibility of the project funding was also regarded as potentially challenging in terms of ensuring that the key objectives were met:

> I was a bit concerned initially, as the project manager, in terms of knowing our target was energy efficiency and carbon reduction … there was the balance to be struck … having community led projects, … but then also having that overarching aim and then, it was like: would we get people who would be interested in some kind of energy related project? Would it only be about food growing, and … that wouldn't really help us to achieve our target.
>
> *(Interview 1, Municipal Authority, London, April 2011)*

The upkeep of the experiment involved a fine balance between investing in those areas which were visible and had support across the community, and those through which the most carbon reduction could be achieved so as to meet targets. In this respect, the CESP retrofit on the Loughborough Estate was fundamental as an investment with long term outcomes. However, because the physical retrofit work did not start until 2012, already some way into the funding period of the LCZ, smaller scale projects such as Draughtbusters underpinned the maintenance of the LCZ in practice.

Alongside the development of tangible projects, maintaining the LCZ also involved an ongoing process of integrating the initiatives and targets into the community. Specific mechanisms for engaging the community were included in the bid (including regular meetings with stakeholders, events and newsletters) and a delivery plan which were designed to raise the profile and impact of the zone and to ensure that it circulated through locally embedded social networks:

> The Brixton LCZ intends to develop ways of engaging socially excluded communities in carbon reduction and sustainable living. Local residents who are already working to improve their communities have been engaged in the project; their contacts and skills will be used to engage further members of the community. Residents will be engaged on a house by house and street by street basis through home visits, events and workshops … These activities will be designed to be enjoyable, educational and practical and to bring residents together.
>
> *(London Borough of Lambeth 2009b: 28)*

An ongoing challenge has been defining a constituency relevant to the climate change agenda. This is partly because the tightly-defined LCZ was seen by some as spatially arbitrary and not readily equal to a community. However, it was also because of the recognition that Brixton is comprised of multiple communities, both geographic communities and communities of interest which are spatially distributed. Those involved in the Brixton LCZ were able to see both the advantages of funding targeted at a specific area, and the fact that such funding fails to adequately recognize that Brixton is complex, with multiple communities in it. This has implications in terms of sustaining community engagement in projects. For example, when TTB first started, there was a focus on participation, but since then the focus on operational activity means that the ability to undertake community involvement has been more limited. Conversely, as a newer group, the Brixton Energy Group was able to move forward very quickly with a great deal of enthusiasm due to the dynamic and diverse group of people involved and the knowledge and expertise within the group, which in turn attracted a wider community interest. Furthermore, the challenge of reaching out to socially excluded groups has been recognized from the outset as one that is time consuming and requires more creative ways to engage residents that focus on building community links and capacities:

> We knew that for it to be effective if you just used normal general communications it wouldn't really penetrate through the community, we wouldn't be able to get them on board. By investing in people skills and looking at the community engagement aspect of things, is the only way we could address a lot of the barriers and get the key messages through.
>
> *(Interview 1, Municipal Authority, London, April 2011)*

Ensuring the circulation of the LCZ and its potential benefits across different communities, there is widespread recognition of the importance of the message coming from within the community itself. For example, one of the successes of the Green Doctors scheme was acting as a mechanism for peer-to-peer engagement. As 'champions' within the community, they were able to engage people and build a more organic response to climate change. Yet, despite these efforts, this has proved more difficult to achieve in practice. For example, respondents noted that the early curtain lining workshops, held under the auspices of the Community Draughtbusters, attracted few people. However, over time the number of participants steadily increased indicating the potential to engage more people. One issue here is the timeframe of the LCZ programme. As one respondent noted, 'there's only a limited amount to which you can actually affect how quickly you network with others' (Interview 2, Municipal Authority, London, May 2011) in relation to both getting information into the community but also developing longer-term relationships.

In this way, the LCZ has come to be embedded within the neighbourhood, although it is difficult to assess at this point the extent to which it has also travelled across wider circulations in the city. The project has contributed to shape London carbon flows, for example, through the implementation of projects such as CESP. Moreover, interventions within the remit of the LCZ have come to redirect flows of knowledge, finance and resources within London. One key example is that of Brixton Energy. The notion of a community-based energy project in Brixton had been developed over a period of years, primarily through the TTB Buildings and Energy group. Through a circulation of resources that was stimulated by the LCZ – initially in terms of staffing, rather than financial support – Brixton Energy was supported to develop as a social enterprise. This intervention in itself has acted to create and redirect resources in a number of ways. Through a Community Energy Efficiency Fund – financed through part of the return to shareholders being redirected – energy efficiency measures going beyond the Loughborough Estate to cover the whole of Brixton are being funded. At the same time, Brixton Energy also acts to connect with and contribute to a wider network – Repowering London – by providing knowledge and resources to community groups beyond Brixton as a way of promoting and facilitate the wide-scale development and local ownership of renewable energy projects across South London. Maintaining the experiment in this way has led to a reconfiguration of wider networks in the city of London.

Living the Low Carbon Zone

While carbon reduction has proven critical in the making and maintenance of Brixton LCZ, we find that as the LCZ comes to be integrated into everyday norms and practices of a heterogeneous community and the LCZ's key partners, carbon itself fades from view. The LCZ has engaged with communities to achieve efficient technical solutions in particular neighbourhoods (e.g. through concentrated retrofitting) and behaviour change. In its design, both objectives imagine the integration of carbon reduction as relatively unproblematic: the notion of development as usual remains uncompromised, with carbon reduction presented as an add-on, in which ambitious targets for emissions reduction are seeing as feasible without fundamentally challenging lifestyles and cultures of energy use. Yet, in practice, bringing together low carbon, fuel poverty and long-standing interests in alternative forms of economic development meant that ideas of low carbon living have been challenged as the LCZ has been embedded in daily practices. Partly, this has been related to a proactive approach to engaging with the community, in which an empowering agenda – beyond simply involving – has been adopted under the premise that the LCZ will be the catalyst for longer-term behaviour change. This raises issues about what 'empowering' actually means, and whether it is related to a broader delegation of emission reductions and self-governing to citizens.

The day-to-day LCZ project work moved away from presenting climate change as a standalone issue, and has instead focused on building capacity within the local community, adopting a holistic approach to issues of social, economic and environmental sustainability. This community-based approach, as we discussed above, has been critical in ensuring that the LCZ is embedded and made visible within the city, but at the same time has displaced carbon to the background. As one respondent noted:

> It is around community cohesion. It is around addressing fuel poverty, bringing homes to a decent standard, it is about reducing carbon emissions because that – in terms of the GLA funding that is the main aim: the project

> aims to bring about reduction of 20.12 per cent in time for March 2012, for the Olympics. That is overarching but ... Brixton Low Carbon zone is a bit more about a holistic approach because ... if you just pitch it, if all we hear about energy and carbon reduction, people would not engage in the whole process. But now with our engagement it's also about giving people a bit more of a broader understanding of it. You can do all you can in terms of improving a building fabric and structure to make it energy efficient, but then if people are not going to understand the value behind that and just continue using high amounts of energy ...
>
> *(Interview 1, Municipal Authority, London, April 2011)*

This embedding of carbon and climate change into the community in an invisible way has partly been a result of the way in which the LCZ was seen as a means of addressing fuel poverty. In showing the potential synergies between low carbon and fuel poverty, Brixton LCZ sought to leverage the potential of each agenda for the other through the capital investment programmes at the Loughborough Estate, as well as in interventions such as the Green Doctors, Brixton Energy and Draughtbusters. For example, residents whose homes had been draughtbusted were more concerned with keeping homes warm than reducing carbon emissions.

Equally, the focus on building social capital and aligning diverse interests meant project designers and implementation officers took existing capacities and interests within the community as their starting point. One longstanding area of community activity within Brixton has been food growing. Engaging with people around this issue has provided a way for the LCZ to start to engage with the community. This initial focus on food and similar activities has – over time – evolved so that it has been possible to include more work on climate change and energy. From this perspective, community activity was seen as engaging people more broadly and allowing people to feel empowered rather than placing specific duties on them for carbon reduction. This in turn has led to the creation of a space within which climate change has found purchase in the community, even if this is not in such direct terms and where the climate change challenge is viewed holistically, encompassing carbon reduction alongside building community capacity and local resilience. The specific discourse of climate change as being about achieving particular emissions reductions targets, while critical to the initiation of the scheme, is one that is largely made invisible through the ongoing work involved in the practice of implementing the Low Carbon Zone.

The 'living' of Brixton LCZ as a climate change experiment has therefore required an engagement with a very open sense of what it might mean to undertake carbon reduction in practice. This has led to the production of low carbon subjects, building on previous experiences of food sustainability. Yet at the same time, the Brixton LCZ was premised on achieving specific changes in behaviour in relation to carbon reduction. This has proven to be particularly challenging, at least in part because of the timescales involved:

FIGURE 10.2 Food growing, Friends of Loughborough Park. Photo © S. Fuller, used with permission.

> Two years is a really short time to be able to say – we're doing a lot of behaviour change work, a lot of work which will have a lasting impact. A lot of the work that we're doing will not show in the figures that are going to come at the end of 2012. But we're having an impact on behaviour, a lifestyle change. For the long term, and how do you evaluate that? How would you be able to capture that?
>
> *(Interview 1, Municipal Authority, London, April 2011)*

As a way of overcoming this challenge, one key aim for those involved in the implementation of the scheme has been to get projects off the ground that can be taken forward by other partners after the end of the funding period:

> We're only here till 2012, as people, and as a resource and the money. But they're the ones who are going to be here longer. So the way I've constantly told them is: our legacy is to be able to support some capacity building and give you as much as we can … most of the time they have so much expertise and skills within themselves, really.
>
> *(Interview 1, Municipal Authority, London, April 2011)*

In practice, there has been a movement from targets for individual behaviour change to collective activities that enable new institutional arrangements around energy. In terms of how this is achieved in practice, developing formal models and social enterprises, such as those that emerged from Draughtbusters and Brixton Energy, is one way of ensuring longevity. This challenges notions about the possibility to create low carbon subjects from an instrumental approach to behaviour and highlights the multiplicity of organic processes that intervene changing broader cultures of energy use and regulate self-governing practices. Here, the process of separating the operation of these activities from Lambeth Council was critical. This transition from funding technical interventions such as those found on the Loughborough Estate to a process that focused on empowering citizens and visioning collective futures may therefore be more powerful as a way of ensuring future low carbon practices with or without the LCZ.

Impacts and implications

All experiments are moulded and shaped through the processes through which they are constituted. Yet Brixton LCZ is a particularly clear example of a project in which the processes of making, maintaining and living have led to the emergence of new logics which both question the assumptions that underpinned the experiment in the first place and have fostered new avenues for experimentation within a given locality. The key aspect of Brixton LCZ is that, while its initial proponents approached it as a technical problem that could be calculated, as it progressed it became clear that the main challenge was to make it compelling enough to align interests and resources around the intervention. To become compelling, a narrow and rather technocratic approach to addressing climate change has been transformed through the engagement of a variety of community interests into a new agenda addressing energy poverty through social enterprise.

The Brixton LCZ was designed to be led by the local authority, in this case the London Borough of Lambeth, and implemented through the involvement of a wide variety of partners. Operationalizing the experiment therefore required both building capacity of these partners and allocating responsibility. This has had implications for the ways in which it was maintained. On the one hand, the local authority remains at the forefront as the key delivery agency, with ultimate responsibility for the delivery of the project. On the other hand, this responsibility is constantly being re-allocated and renegotiated with the community and local partners. This involves complex and ongoing negotiations that lead to specific forms of carbon subjectification, and in the Brixton LCZ move away from self-regulation towards a sense of collective responsibility within Brixton's imagined communities.

Brixton LCZ was made visible both through concrete interventions, especially the CESP funding and the range of associated small-scale interventions that accompanied the development of the experiment, and through an ongoing process of community engagement that will eventually determine the extent to which the experiment is sustainable. However, this work is not amenable to calculation in relation to targets or other performance indicators. Perhaps because of this, carbon reductions have increasingly been placed in the background as the experiment has developed, almost a side-benefit of keeping homes warm rather than the central focus of sustainable energy practices. Whereas the devolution of responsibility to communities appears to question public service values that underlie the role of the local authority, it also opens up opportunities for citizens to access resources and to get involved in interventions that they support. By building on the existing strengths of the community, the experiment has succeeded in bringing about behaviour change strategies that confront citizens with direct challenges to their lifestyles.

Overall, Brixton LCZ highlights that the flexibility to adapt objectives to local concerns, harness available resources and take advantage of ongoing initiatives can be critical in shaping the extent to which experiments have purchase beyond their narrow context of operation – in this case, an arbitrarily defined administrative area. Yet such flexibility is also a factor that reduces the capacity of the experiment to demonstrate the extent to which original fixed objectives for carbon reduction can be attained, particularly as the project moved into interventions less amenable to calculation. For the current mayor, Boris Johnson, 'the time for trials and experimentation is over' as London moves towards mainstreaming low carbon development as part of business as normal (GLA n.d. a). The case of Brixton suggests that there is still much to learn about how, and for whom, experimentation can challenge and sustain such visions.

References

Bulkeley, H. and M. Betsill (2003) *Cities and Climate Change: Urban Sustainability and Global Environmental Governance*. London: Routledge.

Bulkeley, H. and K. Kern (2006) Local government and the governing of climate change in Germany and the UK. *Urban Studies*, 43: 2237–59.

Bulkeley, H. and Schroeder, H. (2008) *Governing Climate Change Post-2012: The Role Of Global Cities*. London: Tyndall Centre for Climate Change Research.

DECC (2011) *Evaluation of the Community Energy Saving Programme*. London: Department of Energy and Climate Change.

GLA (2004) *Green Light to Clean Power: The Mayor's Energy Strategy*. London: Greater London Authority.

——(2007) *Action Today to Protect Tomorrow: The Mayor's Climate Change Action Plan*. London: Greater London Authority.

——(2009) *Mayor Announces £3m to Kickstart 10 'Energy-busting' Low Carbon Zones in the Capital*. Greater London Authority Press Release, 1 May.

——(2011) *Delivering London's Energy Future: The Mayor's Climate Change Mitigation and Energy Strategy*. London: Greater London Authority.

——(n.d. a) *Aiming for Big Reductions in Carbon Emissions*. Available online at http://www.london.gov.uk/priorities/environment/tackling-climate-change (accessed 07 February 2014).

——(n.d. b) *Developing Low Carbon Zones to Help Cut Local Emissions*. Available online at http://www.london.gov.uk/priorities/environment/tackling-climate-change/developing-low-carbon-zones-to-help-cut-local-emissions (accessed 07 February 2014).

Hopkins, R. and Lipman, P. (2009) *Who We Are and What We Do*. Totnes: Transition Network.

London Borough of Lambeth (2009a) *Future Brixton: a Masterplan Strategy for Brixton*. London: Lambeth Council.

——(2009b) *Application for Low Carbon Zone*.
——(2010) *Low Carbon Zone Briefing*, April 2010.
——(2013a) *Explore Lambeth*. Available online at http://www.lambeth.gov.uk/explore-lambeth (accessed 30 June 2014).
——(2013b) *About Brixton*. Available online at http://www.lambeth.gov.uk/explore-lambeth/brixton (accessed 30 June 2104).
Mulugetta, Y., Jackson, T. and Van der Horst, D. (2010) 'Carbon reduction at community scale'. *Energy Policy*, 38: 7541–5.
Thatcher, M. (1988) Speech to the Royal Society, Fishmongers' Hall, City of London, 27 September. Available online at http://www.margaretthatcher.org/document/107346 (accessed 23 January 2014).
Walker, G. (2011) 'The role for "community" in carbon governance'. *Interdisciplinary Reviews Climate Change*, 2: 777–82.

PART III

11

EXPERIMENTATION AND THE POLITICS OF JUSTICE IN URBAN CLIMATE GOVERNANCE

Introduction

In this book, we have developed and advanced a conceptual framework for understanding climate change experimentation. Viewing urban climate governance as conditioned through the pursuit of what Li (2007) has called the 'will to improve', we have shown how climate change experiments have come to be assembled, or made, through processes of problematization, alignment, rendering technical and rendering compelling; how they have (or have not) been maintained through processes of upkeep and metabolic adjustment; and how they have (or have not been) lived through processes of subjectification, contestation and resistance.

Ghosting this narrative, always present yet not until this point brought explicitly into focus, has been the notion that one of the key ways in which experimentation is shaping the urban politics of climate change is by addressing the ways in which climate change articulates with issues of social and environmental justice in the city. This chapter examines how notions of climate justice are forged and contested in the politics of experimentation. We start by examining how the idea of climate justice has been approached to date, contextualizing the predominantly international-scale debates which have been dominant in the literature with more recent interventions from the perspectives of the literature on environmental justice (EJ) and urban political ecology (UPE). These perspectives suggest that, in addition to the key issues of the *responsibilities* to act on climate change and the *rights* to benefit from it usually considered in terms of their distribution across society, we also need to attend to the ways in which climate politics structures the possibilities for participation and procedural justice, and to engage with the idea of *recognition*, which involves accounting for the ways in which the possibilities of accessing rights and responsibilities are already structured within society, and seeking to redress such inequalities.

In the second part of the chapter, we turn to examining the ways in which the making, maintaining and living of climate change experiments appeals to various justice claims in different configurations, whilst also shaping the ways in which climate justice comes to be enacted or resisted. Making experiments, for instance, raises questions of inclusion and exclusion; maintaining experiments depends on particular circulations of the costs and benefits of climate change action within the city; and living experiments casts light on the new rights and responsibilities that are emerging through climate change experimentation, and the new climate subjects in whom these rights and responsibilities are invested. But experiments also highlight – and in many cases seek to rectify or address (as in Monterrey, Philadelphia, Cape Town and São Paulo) – the pre-existing challenges of injustice in the city, challenges that frequently manifest themselves as structural inequalities.

Climate change experiments, we suggest, are new arenas in which urban politics play out, opening up political debate. In this sense, they challenge the view that climate change action fosters what has been called a post-political condition (Swyngedouw 2009, 2011), in which true political debate has been replaced by expert-led, technocratic policy formation and decision-making. Climate change experiments raise these political questions both at the broad scale – in terms of who should intervene, how and with what consequences – and the micro scale, in terms of how climate rights can be accessed and responsibilities exercised by particular actors in particular geographical contexts, and by engaging directly with resistance and contestation. In drawing the threads together, we argue that climate change experimentation is integral to urban politics. Experiments act on the present, but in the name of (multiple) visions of the future. In so doing they are not only the key sites of current urban climate change governance, but also the spaces in which the very meaning of justice in the city is being negotiated, contested and enacted.

From international principles to the urban politics of climate justice

Climate change mitigation has long been seen as a justice issue, predominantly focused on questions of international and inter-generational distributive and procedural justice. In particular, debates have focused on the distribution of rights and responsibilities for emitting greenhouse gases or taking action to mitigate the effects of such emissions. More recently, climate change adaptation has also come to be seen as a justice issue, largely through the lenses of the rights/responsibilities framing developed in relation to mitigation: adaptation rights include the right to protection from climate change, and responsibilities include who might be involved in the provision of compensation for potential or actual harm, for instance through climate finance (Gardiner 2004; Paavola and Adger 2006; Page 2008). Commentators have emphasized that, whilst principles such as common but differentiated responsibility have broad support, operationalizing these principles has been rather more problematic, frequently leading international negotiations into complex territory, such as defining the types of emissions involved, what constitutes essential versus nonessential uses of energy or production of greenhouse gas emissions, and even setting the timeframes over which calculations should be made (Füssel 2010). The situation is even more complex with regard to adaptation, where uneven geographies of both risk and responsibility are overlain by uncertainty about potential effects and their ability to be correctly attributed, at the same time as adaptation measures themselves create new uneven distributions of both burdens and benefits (Bulkeley *et al.* 2013). In an attempt to break the various deadlocks, procedural justice considerations have become particularly important in building legitimacy for policy interventions in the absence of consensus over what the distributions should look like. The outcome is a conceptualization of climate justice as matter of balancing competing rights and responsibilities through a mixture of distributive and procedural mechanisms (Bulkeley *et al.* 2014).

However, as theorists working from EJ and UPE perspectives have argued, justice entails more than achieving a degree of parity in the distribution of burdens and benefits or the use of fair procedures. Drawing on debates from within political philosophy, they have argued that, in addition to distributional and procedural issues, justice involves dealing with issues of recognition (Fraser 1997; Tully 2000; Honneth 2004). Nancy Fraser's (1997) approach to recognition has been particularly influential in EJ theory, which has embraced her argument that socio-economic injustices are fundamentally tied to 'cultural or symbolic' injustices (1997: 14) against certain groups (such as women, the working class, or particular racial or ethnic groups). EJ scholars – following the lead of David Schlosberg (2004) – have thus argued that justice must be seen as trivalent, focusing on issues of distribution, procedure *and* recognition (e.g. Sze and London 2008; Walker 2009; Martin 2013). Arguably, the focus on the urban scale in EJ scholarship underpins this re-theorization of justice, since socio-environmental inequalities are heightened by the proximity of communities, proximity which highlights the structural patterns of inequality and uneven geographies of opportunity, affluence and well-being, as well as the role of environmental factors in reproducing them (Swyngedouw 2004). In terms of climate change experiments, it is in the processes of *making* through which we suggest questions of recognition primarily emerge, because the alignment of multiple actors required to successfully assemble an experiment serves to define and contain the field of intervention in accordance with assumptions concerning what and who can and should be improved in relation to climate change, and practices of calculation serve to reinforce or subvert existing forms of social order.

This new conceptualization of climate justice also changes the debate about rights and responsibilities for climate change mitigation and adaptation at the urban scale, from one focused on what individuals and institutions stand to gain or lose, to one asking questions about the nature of urban politics more broadly. Questions of environmental responsibility are often represented as a matter of attitudes, behaviour and personal choice (Shove 2010). In the field of climate change, this can be seen, for example, in studies that examine the basis for establishing personal carbon budgets (e.g. Roberts and Thumim 2006), or in those that seek to explain why individuals do not perceive climate change as a moral imperative to change their actions (Markowitz and Shariff 2012). However, the new conceptualization of climate justice runs counter to this trend, arguing instead that climate change is best understood as 'a problem of many hands' because many people share in the actions leading to it, and thus an overly individualistic view on responsibility may lead to the idea that nobody is responsible for it (van de Poel *et al.* 2012). This shifts focus away from individual behaviour and individual responsibility towards the view that responsibility for climate change is a collective responsibility (Castán Broto 2013), despite the fact that various groups seek to transfer this responsibility to social institutions, including elected officials, risk managers and central authorities. It builds on an established body of work within EJ which has stressed that environmental responsibility must be understood in terms of existing systems of production and the social relations on which they are built (Agyeman and Evans 2004; Agyeman 2002; Watson and Bulkeley 2005). At the same time, the attribution of responsibilities is closely linked to the processes of knowledge production, and who has the capacity to enact particular types of knowledge about what constitutes a response (Castán Broto 2012). This brings us to the way experiments are *maintained*, because fundamental to maintenance is the ability to create responsibilities and assign them to actors and to continually reinforce such configurations, and also how experiments are *lived*, which requires the formation of subjectivities through which responsibilities for being low carbon are enacted and contested.

A similar change in the approach to climate rights results from the new conceptualization of climate justice. Rather than thinking of rights in individual terms, the addition of recognition to a trivalent conception of justice requires paying attention to how the politics of experimentation configures rights in relation to the wider socio-material milieu. Lefevbre famously called for a right to the city and, notwithstanding the ongoing debates about the significance and indeed actual meaning of this slogan, it captures concisely the sense that justice is not merely about creating a better city for its inhabitants, but also the right of those inhabitants to decide on what a better city means (Harvey 2003; Brenner *et al.* 2012). In short, this is a call for an urban politics of rights and, within this debate, the right to the city is constructed as a moral claim to:

> multiple rights that are incorporated here: not just one, not just a right to public space, or a right to information and transparency in government, or a right to access to the centre, or a right to this service or that, but the right to a totality, a complexity, in which each of the parts of a single whole, to which the right is demanded.
>
> *(Marcuse 2012: 35)*

From this perspective, ensuring urban justice is not about access to specific individual rights, but rather a process that exposes, challenges and reconfigures the urban patterns of unequal access to resources (Marcuse 2012). In terms of climate change experiments, the processes of upkeep and adjustment which reconfigure and circulate experiments as they are *maintained* become critical as the means through which rights to benefit from urban responses to climate change – whether this be through the redirection of benefits or protection from harm – are realized. At the same time, realizing such rights requires that new forms of subjectivity and everyday practice are enacted in which climate rights in the city can be *lived*.

Rather than being a matter of the execution of well-considered principles, we find that climate change experimentation involves a politics of climate justice in which the processes of making, maintaining and living serve not only to frame climate change in terms of responsibilities, rights and recognition, but also facilitate the realization of climate justice in these terms. Recognition is most closely tied to the *making* of experiments, particularly in terms of processes of alignment and calculation, though we also suggest that the politics of legitimacy and inclusion are also manifest in the way experiments come to be made compelling through powers of seduction and persuasion (Allen 2003). The *maintaining* of experiments involves both the upkeep of these forms of recognition and the adjustment of metabolic circulations, such that both rights to access resources or be protected from harm are reconfigured, and responsibilities for action shifted according to the new urban landscape created through the insertion of the experiment (see Chapter 12). The *living* of experiments involves the translation of the principles and material shifts invoked through the work of experimentation into distinct forms of subjectivities and everyday practices within which responsibilities and rights come to be sustained and resisted (Table 11.1).

In the next section, we examine how justice finds expression in and is articulated through climate change experiments, drawing on the case studies presented in the previous chapters. Table 11.1 highlighted some of the ways in which each of the elements of climate justice relate to these experimental phases, and in the next section we build on this, arguing that experimentation both serves to construct and enact a politics of climate justice, which unfolds through the processes of making, maintaining and living experiments.

TABLE 11.1 The politics of climate justice in urban climate change experimentation

Principles of climate justice	*Processes of experimentation*
Recognition	• Making: alignment and calculation • Maintaining: reinforced through upkeep • Living: reinforced through subjectivities
Rights	• Maintaining: metabolic adjustment reconfigures access to resources/protection • Living: subjectivities and everyday practice central to realisation of climate rights
Responsibilities	• Maintaining: metabolic adjustment serves to rework landscape of responsibilities for response • Living: subjectivities and everyday practice central to realising both the prudential and ethical dimensions of responsibilities

Locating the politics of justice in climate change experiments

Treating climate justice as having three dimensions – rights, responsibilities and recognition – in this section we examine how each is constituted across the making, maintaining and living of climate change experimentation. This both casts light on the multiplicity of justice claims being engaged by experiments through these processes, and helps us give greater form to the interpretations and meanings of rights, responsibilities and recognition that are emerging through the process of experimentation. For instance, rights can be limited, open or temporally constrained (e.g. future rights). Responsibility, likewise, can be individual, collective, assigned to specific actors or institutions or even avoided. Finally, recognition can be complete, partial, missing, deliberate, inadvertent, productive or even counterproductive. All of these facets of justice are also articulated in distributive or procedural formulations in ways that might be spatially or temporally contingent: top-down or bottom-up; centralized; decentralized; or even polycentric. By reflecting on how processes of making, maintaining and living have led to particular forms of justice discourse and action across the eight experiments we have examined in this book, we show how experiments are embedded within the broader urban politics of justice at the same time as they are shaped by and shape the international politics of climate justice.

Making experiments and the conception of climate justice

The making of experiments involves processes of problematization, alignment, rendering technical and rendering compelling, none of which is neutral in justice terms. In fact, the very act of formulating a problem and positioning an experiment as a solution to it is loaded with value judgements. For example, in Monterrery, ViDA explicitly linked discourses of urban justice with discourses of climate action, by actively positioning climate change as a reason for constructing low cost, affordable housing for those unable to access it privately. ViDA leveraged climate change towards reducing levels of housing inequality, which are so stark in Monterrey. By contrast, a conscious attention to justice was entirely absent from the processes of problematization at T-Zed, and this carried through in the other processes of making the experiment. T-Zed's goal of creating a low carbon gated community targeted at the aspirational middle class meant that the project served to reproduce existing structural inequalities within the city, suggesting that it is those with access to money who are valid beneficiaries of climate change action. The making of the experiment thus ascribed climate rights to the middle-class residents who bought into the development, while the discourse of self-sufficiency which was dominant in the development simultaneously circumscribed the rights of nearby (poorer) residents who might have hoped to lay claim to benefiting from the water collection structures constructed as part of the development.

Considerations of environmental justice entail the recognition that climate change is entangled with ongoing social issues of energy poverty, urban health and access to service provision, and how co-benefits of climate responses should be constructed. Climate change thus becomes compelling in part because of the ways in which it can be aligned with other issues. In Cape Town, London, São Paulo, Monterrey and Philadelphia, alignments were forged between climate change and issues of disadvantage and structural inequality. In Cape Town, the Mamre project sought to mobilize carbon finance to address the demands of the poor and improve local health, and in São Paulo specific alignments between energy and climate change led to the idea that shifting to renewably-powered hot water systems was a way to address poverty and disadvantage. In London, the LCZ experiment in Brixton challenged the boundaries of who was enrolled in responding to the issue, creating new justice demands by aligning climate change with pre-existing concerns about the existence and extent of energy poverty and the geographically uneven nature of this urban inequality. In Monterrey, the ongoing challenges of poverty were recognized at the local level, but the project failed to reproduce this attention to the recognition dimension of justice as it circulated at the national level, perhaps because of the strong demands to provide housing at least cost, paired with the need to create more transferable financial instruments in order to achieve this in the form of the Green Mortgage. In other cases, such as Bangalore and Berlin, issues of disadvantage were much less prominent, and other logics, such as carbon control or urban ecological security, provided the compelling reason to act. Importantly, in justice terms, these tended to prioritize the visions of urban elites over others, and both of these experiments failed to engage substantively with the recognition dimension of justice.

In some experiments, justice considerations were central to forging alignment between the divergent interests of actors in the city. In ViDA, for example, it was the inclusion of climate change considerations in social housing that precipitated the access to financial support in the current context of housing provision in the country. The case is similar to those of São Paulo and Cape Town, where climate change initiatives – solar water heaters, ceilings – added value to expansive programmes of social housing. The question is whether recognition of the alignment between social housing and climate change interests positions these experiments as advancing a specific programme of insertion of the poor in dominant housing markets, or whether it serves to constitute truly emancipatory alternatives. While the case of Mexico, and the growing industry of Infonavit, demonstrates how climate change arguments can be appropriated to support the interests of a voracious housing industry, in Brazil and Cape Town the experiments have generated debates about what kind of housing is possible and desirable, uncovering the tensions between providing houses on a large scale and providing houses of quality for urban dwellers which will be sustainable over the long term. In Philadelphia, the Coolest Block contest attempted to recognize the potential of communities to engage in refurbishing the built environment, but the integration of the measures in wider capital markets has redirected climate change action towards those urban citizens who can afford loans, and thus the extent to which social concerns are at the heart of the initiative is largely unproved.

The process of making is one of assembling resources, but also assembling narratives and visions that can give impetus to the processes that generate the experiment. A key aspect of this process involves rendering the problem at hand technical. This form of calculation always involves defining who is and is not included in the design of the experiment, for whom the experiment is intended to act, and who is considered a valid beneficiary of the experiment's actions. This raises the question of recognition in multiple dimensions. For example, Berlin's Solar Atlas positioned business/investors as valid beneficiaries of increasing the amount of renewable energy generation in the city, and home-owners rather than renters as valid beneficiaries of the action. This shaped the experiment, both in terms of its target constituency and what it was and was not ultimately able to achieve. In addition, processes of alignment require strategic selection of key partners, which establishes certain actors as valid participants. So both Philadelphia and Hong Kong show that certain ways of engaging with low carbon exclude particular actors and communities, even where the intent is inclusion. This can have both positive and negative effects in recognition terms. For instance, the explicit focus of Transition Brixton is the locality of Brixton in South London. This focus inherently excludes areas outside Brixton, but it does so in order to counter a prevailing lack of recognition of Brixton. In Philadelphia, by contrast, the focus on one successful block came at the expense of sustained engagement with a myriad of other blocks in equal need of renovation. Here, the demands of making the experiment compelling necessitate creating a narrative of a successful experiment in order to seduce and attract others to try similar projects elsewhere or upscale them. Such processes, through privileging success narratives, tend (as in the case of Philadelphia and Hong Kong) to exclude the visions or needs of those without power, reinforcing existing recognition-based injustices.

Overall, actors struggled between the drive to put forward a normative view on climate change action and the need to recognize the multiple forms of knowledge in relation to different beliefs and values about what kind of action is desirable. This is perhaps most prominent in the Solar Atlas experiment, in which an overzealous concern about visual representation and cutting-edge technology served to neglect more accessible forms of knowledge about how to plan for solar energy in the city. While focusing on a particular form of knowledge may be necessary at the delicate stage in which making the experiment calculable and compelling requires building a coherent discourse around it, its shortcomings in terms of addressing recognition become evident in relation to the stages of maintaining and living, in which processes of upkeep and subjectification may be necessary to ensure the carefully-constructed narrative of the experiment is integrated within the urban landscape. In the cases of Philadelphia and London, for example, a certain flexibility in the formulation of the project has facilitated a redefinition of the starting terms of reference and their integration within community discourses to the extent which, particularly in London, communities themselves have taken the experiment as an extension of their ongoing work, and have thus further incorporated it into their sustainability practices: making experiments remains an ongoing process.

Maintaining climate experiments and the (re)configuration of climate justice

Maintaining experiments entails processes of upkeep and metabolic transformation, which on the one hand act to one end – ensuring that the experiment is embedded into the normal fabric of life, and becomes in some sense 'unexceptional' within the city – while simultaneously facilitating the circulation of the experiment and its components, to support is reproduction and sustainability over time. Experiments play a dual role here, simultaneously casting light on the circulations of costs and benefits, advantage and disadvantage of the urban contexts in which they are located, by virtue of their distinctness, yet at the same time they are dependent on these existing circulations and socio-technical orders to acquire legitimacy. As a result, experiments simultaneously sustain and challenge existing patterns of injustice and inequality in the city.

The T-Zed project demonstrates this. Whilst the making of the experiment was strongly focused on climate rights, and particularly the rights of what were considered to be worthy middle-class residents to quality of life as well as the benefits of low carbon homes, as residents moved into the development it became increasingly clear that these rights were not without responsibilities, at least in the eyes of the developers. Maintaining T-Zed was thus characterized by a series of attempts to regulate behaviour by creating a sense of responsibility in residents, or what we term a low carbon disposition (Chapter 2). For instance, the developers mandated the use of energy-saving devices, such as preinstalled and tailored refrigerators, which residents found inconvenient, and installed smart meters in the homes they constructed, actively referring to them as conscience meters. In this sense, the upkeep of the experiment created a very different justice dynamic through its maintenance to that forged through its making. Yet T-Zed also stood as an ongoing example of how common pool resources in Bangalore and more broadly in India are being encroached upon by private developers whose clientele are overwhelmingly the upwardly-mobile middle classes and, while seeking to intervene at the level of individual responsibilities through the conscience meters and fridges, T-Zed ultimately did not question the unsustainable patterns of urban development that justified the development. For example, T-Zed implicitly prescribed rights on the middle-class residents to a supply of safe water which was not extended to the urban poor in the immediate vicinity. The case of T-Zed shows how questions about climate change and resource access rights can only be apprehended in relation to the wider urban issues of development in Bangalore. The construction of new developments is affecting both agricultural practices in the fringe of the city and natural habitats, both which are considered of being of less value. But, most importantly – and this is most obvious in relation to the metabolic adjustment of the experiment – all these developments, no matter how green, depend on the excavation of boreholes and the encroachment of the scarce water resources of the city. As water becomes the paramount issue for the future of the city, these justice issues are further woven into the rights of future generations and the capacity of the city to sustain itself over the long term.

Thus, maintenance emerges as a set of processes connected to the reinforcement or reallocation of rights and responsibilities, and the extent to which this reproduces or challenges existing injustices cannot be inferred alone from the specific ways in which the experiment is assembled. In São Paulo, while the solar hot water systems were seen as a way of saving both money and energy for poor residents, they cemented the idea that access to hot water is a right, and that showers should be available to all. The circulation of technology has thus served as a means to reconfigure rights to access hot water. Yet the type of energy and its mode of consumption is restricted to hot water, without necessarily challenging existing arrangements of energy provision through other centralized systems. When emerging in the margins, experiments may tend to ratify existing rights configurations by limiting the extent to which any form of reconfiguration is possible, by restricting the technological and social functions of any given experimental assemblage. Such limitations are also visible in the ViDA experiment in Monterrey, which was built on the notion that people have a right to own their own homes. However, such a right is conceptualized as a right to have the same home, that is, a home that is virtually the same as the next one, as they are all made with the same template. While this responds to the needs of reducing costs to entice developers to test new bioclimatic designs, it simultaneously redefines the idea of home and the extent to which residents are entitled to adapt, personalise or modify it to make it distinct. The eventual normalization of the experiment through practices of everyday maintenance took place as residents sought to redefine the terms in which they access housing by modifying the house to fit their daily practices. In so doing, the homogenization paradigm underlying the project and, more broadly, the housing industry in Mexico concerning how housing resources should be distributed were challenged. In all these cases, it should be noted, the work of experimental maintenance has the effect of normalizing claims, transforming them from possibilities into assumptions, and thereby foreclosing debate about their implications in broader terms. In doing so, experiments redefine what constitutes acceptable urban development.

Experiments may also serve to highlight how existing urban circulations work against certain actors. In both Philadelphia and Berlin, private interests were positioned as valid beneficiaries of climate change action, and indeed in the final calculation it was such interests that derived the *most* benefit from both of these experiments, at the expense of the urban poor who appeared to remain on the margins. The Coolest Block contest, for example, shows how private interests are positioned as valid beneficiaries of climate change action. Equally, the Solar Atlas in Berlin emphasized private interests not just as valid, but as those who could bring forward and accelerate climate action. In both cases, however, the failure to integrate the experiment within a broader spectrum of social interests has led to stagnation and a lack of circulation of both technological and social innovation.

The case of the Solar Atlas is particularly salient in that it emphasizes that overlooking local housing interests that could shape future policies prevented its broad adoption within the city. Yet the lack of circulation cannot be attributed to this factor alone. The very needs for maintenance predetermine who can access certain information and, hence, how rights for such information can be allocated. We find that the issue of rights is important in two ways. First, whose rights are recognized will help to establish the calculative procedures and rhetoric of the experiments. Some of these rights may thus be already present during the making stage, in terms of who has access to certain resources and capacity, and whether they feel compelled to use these to advance climate change objectives. However, it is through the processes of maintaining experiments that rights come to the fore, especially in terms of metabolic circulations, as the integration of the experiment within the overall fabric of the city is likely to raise underlying tensions and differences in access to resources and services over which the experiment is constructed. Moreover, experiments may help to constitute new rights, to make them visible, and thus reinforce them. In Mamre, for example, the construction of ceilings in poor dwellings is a measure that affects directly the bacterial circulations, the transmission of disease, and hence the urban health of the communities affected. This makes explicit the right to be kept free of disease, even though the policy may be directed towards addressing climate change.

Living experiments and the process of subjectification

Living experiments is fundamentally about creating new subjects relevant to climate change, the process of subjectification. This is of course deeply intertwined with processes of contestation and resistance which shape these subjects. Subjectification is inherently about justice, since the creation of subjects entails bestowing on them rights and also responsibilities which are relevant to their new role with respect to climate change. Climate change experiments are therefore creating both new rights and responsibilities, and new subjects in whom these rights and responsibilities are invested. Subjectification involves citizens and populations regulating themselves – that is, taking action toward solving collective problems – which shifts responsibility for addressing climate change from institutions (government, international agreements, NGOs) towards citizens. In some cases this leads to the re-representation of the consumer as a climate change good citizen (e.g. Bangalore, Hong Kong), but in other cases it also represents challenging understandings of disadvantaged people as being unable to take responsibility for collective problems (Monterrey, São Paulo, Mamre).

The question of subjectification through the living of experiments highlights the extent to which those who participate do so within an acceptable degree of freedom. This refers not just to the actual participation in the experiment but also the extent to which those who will be subject to experimentation can actively participate in the definition, assemblage and calculation that are involved in the process of making and the circulatory practices of maintenance: the extent to which they have a right to the experiment. The connections between making and maintaining experiments and living them is not trivial. Both in Bangalore and Berlin, the experiments struggle to reconcile the living practices idealized by those who made them and the actual patterns of living of those who are subject to it. While in Bangalore residents actually contested the capacity of BCIL, the developer, to prescribe how to live in the development, in Berlin the failure to enrol potential subjects in the assemblage led to the experiment ultimately not being realized. The failure of the Solar Atlas has to be read, as well, in relation to broader patterns of structural inequality in the city that further determine the potential to engage with such innovations. In Philadelphia, the structure of the Coolest Block contest was meant to facilitate the participation of residents, or at least their community leaders, in the active process of making the experiment, but the degree to which citizens could actually intervene in the definition of the intervention was limited by the terms of the context, the perceived favoured blocks and the extent to which certain technologies and strategies were prescribed in advance.

It is perhaps in the cases of LCZ Brixton, London, and Climateers, Hong Kong, where the process of subjectification has been more advanced, and in both cases this can be related to the fact that in both experiments those living the experiment have been allowed to participate in its making and maintenance. In London this has been done through a certain flexibility to engage with ongoing community practices, so that LCZ Brixton came to support established sustainability strategies, hence facilitating the incorporation of new patterns of subjectification with climate change at their centre. In Hong Kong, the emphasis was on creating individual responsibility through individual low carbon strategies, thus putting potential low carbon subjects in charge of the low carbon innovation process. Overall, experiments reconfigure patterns of responsibilities through establishing alternative forms of everyday practice through which daily life is conducted. In Hong Kong, the systematization of everyday practices, their measurement, comparison and competition, as well as further communication through social media, created ideal means through which voluntarily-adopted responsibilities were given value and prominence. Moreover, the process led to its self-reproduction. Once an activity was completed and communicated, low carbon subjects both inspired other action and internalized the action into their daily practices through their repetition.

Those experiments in which concerns with social justice were explicit (e.g. Monterrey, Cape Town, São Paulo, London) may have helped to create new rights for the urban poor, from housing to health to hot water, but they have also advanced the displacement of responsibilities towards poorer groups of population. Both in Monterrey and São Paulo there is a discourse of the potential to address climate change emissions reductions through intervening in the everyday practices of the poorest. Such an outlook may be surprising if we consider the enormous differences in carbon emissions that exist across such cities, and need to be explained in relation to the greater capacity of these governments to intervene in the lives of the urban poor than in those of the privileged. However, they also point to a broader issue in relation to climate change mitigation and the extent to which climate change responsibilities can be extended to all the individuals across the city. In Cape Town, in contrast, responsibilities are not necessarily displaced to residents as much as to the technicians who intervene in the particular form of innovation, so that making experimental interventions for climate change with co-benefits for the urban poor does not automatically prescribe to the urban poor the responsibility for addressing climate change.

Yet, whichever way they are precisely configured, experiments constitute new patterns of responsibility, particularly through the constitution of subjects that have to respond to climate change. Responsibilities may be allocated in relation to everyday practices or in relation to professional practices. The later, unlike the former, frames responsibility not just in terms of the private lives of individuals and the way the experiment becomes a part of everyday landscapes, but rather in relation to the differential capacity to intervene in the urban milieu between elites and other actors, and points to the systematic nature of the response required and the potential for imagining alternative futures.

Conclusion: a new politics of climate justice in the city?

Bringing justice into focus implies reflecting upon what would constitute a new politics of climate justice in the city. In so doing, our analysis begins with a concern with recognition as a central component of climate justice. This means both recognition of needs and potential, but also recognition of the manner in which interventions shape the urban environment. Urban politics are inextricably caught up in the politics of climate justice, in part at least because the politics and practice of climate change experiments are shaped by everyday contestations over the meaning of justice. Our analysis throws light on the constitution of the three dimensions of climate justice, rights, recognition and responsibility, through analysing how such dimensions are deployed in the making, maintaining and living of climate change experiments.

Thus, climate change experiments are integral to both urban politics and the politics of climate justice in several ways. First, through climate change experiments, the roles of traditional actors in society (including the state, civil society and capital) are being rearranged, raising the question of who is and should be ultimately responsible for climate change action, and what the implications of this changing landscape of responsibility might entail. Second, climate change experiments are transforming the patterns of advantage and disadvantage as they are rendered through the physical landscape of the city. This raises the question of power: how it is constituted, extended and circumscribed through the art of experimentation. Finally, climate change experiments are creating new justice challenges, which are in many cases just as pressing as the challenges being thrown up by the changing climate itself. This raises the question of compromise (cf. Li 2007), and particularly how the winners and losers inevitably created through experimentation are to be negotiated. Together, these questions of responsibility, power and compromise raised by climate change experiments are central to the politics of climate justice in the city. Climate change experiments, in other words, are key sites in which the meaning of justice in the city is being tested, contested and negotiated: it is becoming a critical arena for enacting climate justice.

In this context, a key concern is how justice claims themselves may play an important role in assembling experiments and making them compelling, particularly by aligning climate change with other co-benefits and proposals to intervene in order to address a variety of urban challenges. The extent to which the experiment succeeds in bringing about such arguments also determines whether they are reproduced or reassembled in new incarnations, such as in Mexico where the original bioclimatic concerns of ViDA were redefined when the Green Mortgage programme was rolled out nationally. At the same time, experiments also draw attention to how justice claims are embedded in the production of knowledge, and hence how the production of knowledge itself is a political act. As unsanctioned, unregulated and unruly spaces, climate change experiments both provide the opportunity for negotiations over justice within their boundaries and also highlight the pre-existing challenges of urban justice which climate change is increasingly being mobilized to address. In the context of recent assertions that climate governance represents the supreme example of the post-political neoliberal consensus-based mode of governance (Swyngedouw 2009, 2011), climate change experiments exist as reminders of the persistence of politics in shaping urban governance. In this context, climate change experiments have become key sites in which broader contestations about the desired form of the city can be enacted, whether or not the experiment itself consciously mobilizes a justice agenda, and the potential for a politics of urban transformation (Chapter 12).

References

Agyeman, J. (2002) 'Constructing environmental (in)justice: transatlantic tales'. *Environmental Politics*, 11(3): 31–53.

Agyeman, J. and Evans, B. (2004) '"Just sustainability": the emerging discourse of environmental justice in Britain?'. *Geographical Journal*, 170: 155–64.

Allen, J. (2003) *Lost Geographies of Power*. Oxford: Blackwell.

Brenner, N., Marcuse, P. and Mayer, M. (2012) *Cities for People Not for Profit*. London: Routledge.

Bulkeley, H., Carmin, J., Castán Broto, V., Edwards, G. A. S. and Fuller, S. (2013) 'Climate justice and global cities: mapping the emerging discourses'. *Global Environmental Change*, 23(5): 914–25.

Bulkeley, H., Edwards, G. A. S. and Fuller, S. (2014) 'Contesting climate justice in the city: examining politics and practice in urban climate change experiments'. *Global Environmental Change*, 24: 31–40.

Castán Broto, V. (2012) 'Exploring the lay/expert divide: the attribution of responsibilities for coal ash pollution in Tuzla, Bosnia and Herzegovina'. *Local Environment*, 17(8): 879–95.

——(2013) 'Who has moral responsibility for climate change?'. *e-International Relations*: March 6.

Fraser, N. (1997) *Justice Interruptus: Critical Reflections on the 'Postsocialist' Condition*. New York: Routledge.
Füssel, H.-M. (2010) 'How inequitable is the global distribution of responsibility, capability, and vulnerability to climate change: a comprehensive indicator-based assessment'. *Global Environmental Change*, 20(4): 597–611.
Gardiner, S. M. (2004) 'Ethics and global climate change'. *Ethics*, 114(3): 555–600.
Harvey, D. (2003) 'The right to the city'. *International Journal of Urban and Regional Research*, 27(4): 939–41.
Honneth, A. (2004) 'Recognition and justice: outline of a plural theory of justice'. *Acta Sociologica*, 47(4): 351–64.
Li, T. M. (2007) *The Will to Improve: Governmentality, Development, and the Practice of Politics*. Durham, NC: Duke University Press.
Marcuse, P. (2012) 'Whose right(s) to what city?'. In Brenner, N., Marcuse, P. and Mayer, M. (eds) *Cities for People Not for Profit*. London: Routledge, pp. 24–41.
Markowitz, E. M. and Shariff, A. F. (2012) 'Climate change and moral judgement'. *Nature Climate Change*, 2(4): 243–7.
Martin, A. (2013) 'Global environmental in/justice, in practice: introduction'. *The Geographical Journal*, 179(2): 98–104.
Paavola, J. and Adger, W. N. (2006) 'Fair adaptation to climate change'. *Ecological Economics*, 56(4): 594–609.
Page, E. A. (2008) 'Distributing the burdens of climate change'. *Environmental Politics*, 17(4): 556–75.
Roberts, S. and Thumim, J. (2006) *A Rough Guide to Individual Carbon Trading: The Ideas, the Issues and the Next Steps*. UK: Centre for Sustainable Energy.
Schlosberg, D. (2004) 'Reconceiving environmental justice: global movements and political theories'. *Environmental Politics*, 13(3): 517–40.
Shove, E. (2010) 'Beyond the ABC: climate change policy and theories of social change'. *Environment and Planning A*, 42(6): 1273–85.
Swyngedouw, E. (2004) *Social Power and the Urbanization of Water: Flows of Power*. New York: Oxford University Press.
——(2009) 'The antinomies of the postpolitical city: in search of a democratic politics of environmental production'. *International Journal of Urban and Regional Research*, 33(3): 601–20.
——(2011) 'Depoliticized environments: the end of nature, climate change and the post-political condition'. *Royal Institute of Philosophy Supplement*, 69: 253–74.
Sze, J. and London, J. K. (2008) 'Environmental justice at the crossroad'. *Sociology Compass*, 2(4): 1331–54.
Tully, J. (2000) 'Struggles over recognition and distribution'. *Constellations*, 7(4): 469–82.
van de Poel, I., Fahlquist, J. N., Doorn, N., Zwart, S. and Royakkers, L. (2012) 'The problem of many hands: climate change as an example'. *Science and Engineering Ethics*, 18(1): 49–67.
Walker, G. (2009) 'Beyond distribution and proximity: exploring the multiple spatialities of environmental justice'. *Antipode*, 41(4): 614–36.
Watson, M. and Bulkeley, H. (2005) 'Just waste? Municipal waste management and the politics of environmental justice'. *Local Environment*, 10(4): 411–26.

12

URBAN TRANSFORMATION AND URBAN POLITICS

Introduction

In this book, we have argued that understanding the urban politics of climate change requires examining the ways in which the socio-technical and socio-ecological configurations that make up the city are being reordered and aligned with new forms of climate change governmentality. As we argued in Chapter 11, at the heart of any such analysis must be an engagement with the consequences of these processes for social and environmental justice. We start from the argument that a will to improve the urban in relation to climate change has become apparent not only in the policies and strategies of municipalities, but through a range of forms of experimentation undertaken by various actors in multiple guises. Rather than regarding experiments as standalone exemplars, we suggest they provide a key means through which the urban governing of climate change is conducted and accomplished; a form of 'governing by experiment' that enables a response to the uncertainty of climate change, works across the fragmented authority to govern and takes place without overt challenge to existing political and economic structures.

Yet many scholars regard experiments merely as curiosities – interesting to look at but of limited value in the context of the wholesale shift in infrastructure and economic systems deemed to be required to address climate change. Transitions, 'major technological transformations in the way societal functions such as transportation, communication, housing, feeding are fulfilled' (Geels 2002: 1257) have thus been championed as offering the means through which to combat the challenges of decarbonization. Scholarship on socio-technical transitions has sought to understand both how such transformations have happened in the past (e.g. Verbong *et al.* 2008; Geels and Raven 2006), as well as how they might be managed in the future, in order to move society on to a more sustainable footing (e.g. Elzen and Wieczorek 2005; Raven *et al.* 2009; Smith *et al.* 2005). The notion of a low carbon transition has become increasingly popular, yet understandings of what transition might entail vary significantly. Government reports such as the *UK Low Carbon Transition Plan* use the idea of transition to suggest that a rapid systemic change, along a well-marked pathway, can be brought about with minimal impact on quality of life and economic markets (DECC 2009). Transition is directed and purposeful, as government intervenes to move society from our present unsustainable state to a future, more desirable and stable, sustainable moment. In contrast, movements such as Transition Towns have reframed transition as a local issue addressed through community mobilization (Bailey *et al.* 2010). Despite their various differences, transition discourses have in common the sense that their realization depends on identifying clear goals and pathways through which to channel social change.

In this concluding chapter, we argue that the transformative potential of experimentation can be considered in a different light. Our analysis places experimentation in the ongoing, unfolding and heterogeneous set of processes through which the will to improve is pursued. Conceiving of transformation as a transition between two states, in which the first one is putatively less sustainable than the second one, appears rather redundant. Instead, transformation might more productively be considered to involve the continual reconfiguration and renegotiation of socio-technical networks. Transformation is not here an end point, for which such interventions provide the initial signpost, but rather is already present in the politics and practice of governing by experiment. Thus, we examine how processes of transformation are engendered and unfold through experimentation, and the ways in which such purposive interventions can lead to multiple and often unexpected outcomes. Through this approach, we argue that understanding urban transitions, and their politics, requires engagement with a kaleidoscope of plural socio-technical regimes that go to make up the urban, and in which climate change experiments provide critical junctures through which new configurations are assembled, mobilized, normalized and contested (Bulkeley *et al.* 2013).

In the next section, we problematize the geography of low carbon transitions by situating it, explicitly, within the urban contexts in which socio-technical relations of both production and consumption become materialized. In light of this critique, the third section of the chapter develops a new approach, what we term a *topology of shifts*, to signify the processes of change in socio-technical and socio-ecological systems that occur through a myriad of movements, and what this means in turn for conceptualising urban transitions in relation to their making, maintaining and living (Chapter 2). To support our argument, and develop it empirically, in the fourth section we examine the evidence from across the case studies concerning the transformative potential of experiments and how this is realized through different urban milieux. In conclusion, we return to consider the forms of politics to which the art of experimentation is giving rise.

Cities, experimentation and low carbon transitions

A great deal of research has sought to understand how and why innovations and interventions come to be widely taken up. While, for the most part, these literatures, focused on technological innovations and system change, have tended to disregard the urban arena, there has been growing interest in the role that cities might play in transitions for sustainability. Here, we consider how conceiving of the urban as a space for experimentation has been addressed in these debates, and the ways in which current approaches have sought to analyse the processes through which experiments come to be scaled up so as to effect wider system change.

Making space for the urban

For proponents of socio-technical regimes and analysts of transition, the urban has traditionally been excluded from view. Systems and regimes are, for the most part, conceived in national terms, such that they are assumed to operate in geographically even ways with cities regarded as simply locations across which such dynamics play out. More recent debate has, however, sparked a wave of interest in the geographies of transitions (Coenen *et al.* 2012; Truffer and Coenen 2012), in which cities are increasingly positioned as significant (Hodson and Marvin 2010). Geels (2010) identifies two main roles that cities might play in socio-technical transitions: first, the city is conceived as an *actor*, such that city governments may lead transitions; second, it is regarded as a *theatre for action*, providing spaces of innovation that can act as seedbeds for transitions. Our analysis suggests that such approaches may have limited traction in understanding the ways in which the urban figures in transitions. From an actor perspective, we find multiple agents engaged in experimentation, and that agency is not only located in organizations and individuals but socio-materially constituted. Equally, while the notion of the city as a theatre of action has strong roots within urban theory (e.g. Mumford 1997 [1937]: 85), the idea of city as a bounded space where human energies concentrate downplays the generative aspects of urban life (Hubbard 2006) and the multiplicity of connections and alignments of which the urban is comprised (Graham and Healey 1999). A new wave of research has sought instead to consider cities as central in the orchestration of transitions. Adopting a view that takes account of the diverse ways in which the urban is constituted immediately reveals the diverse actors working at multiple scales that form networks and constellations of actors that influence transitions (Carvalho *et al.* 2012; Späth and Rohracher 2012). The presence of such multi-scalar networks points to the importance of considering the ways in which urban governance processes are already enmeshed with purposive efforts at transition (Loorbach and Rotmans 2010). Hodson and Marvin (2012) have shown how different actors in the city have sought to develop transition strategies and the ways in which such strategic interventions circumscribe who partakes in transition processes. Rather than the city forming an actor that can coherently pursue a transition, such processes reveal the contested nature of transitions and the multiple positions occupied by urban actors in these processes (Hodson and Marvin 2012).

Despite the promising advances made in such accounts, they have tended to remain focused on actors and institutions as the means through which transition processes are generated, albeit challenging any simple geographical basis to such forms of governance and pointing to the complex and contested ways in which they unfold. As a result, the obduracy of urban socio-technical systems often appears neglected in such accounts. Obduracy refers to the stubborn persistence of specific socio-technical configurations despite their continuous exposure to change. In her detailed study of obduracy in urban infrastructure, Hommels (2005) suggests that obduracy emerges through three processes. First, according to the theory of technological frames, infrastructure configurations obdurate because different social actors operate within constrained ways of thinking and persistent patterns of interaction. Second, according to the notion of embeddedness, infrastructure is a system of relations which mean that changing one component requires multiple changes in associated elements, in turn meaning that maintaining old configurations is often more feasible than introducing innovations. Finally, the theory of persistent traditions suggests that there are persistent ways of doing things which determine the possible histories of infrastructure along predetermined paths or trajectories. Rather than offering competing explanations, Hommels (2008) suggests that they point to the three-dimensional character of infrastructure obduracy, grounded simultaneously in interactions with, relations between, and endurance of the different components of infrastructure networks. Understanding urban transitions requires, therefore, a perspective that can seek to understand how such forms of obduracy become disrupted and reordered, such that innovations can take place through changing both the idea of what urban infrastructure is, what and who it is for, as well as through shifts in the assemblage and reproduction of urban systems.

In seeking to develop an account of the nature of urban transformations wrought through experimentation that can address a more distributed notion of agency and engage the notion of obduracy, we find it productive to revisit Hughes' (1983) account of the development of large technical systems. In this account, Hughes draws attention to the ways in which such socio-technical networks are contextually, geographically and historically produced, such that, in his case, some cities came to occupy the centre of the transition to electrification while others were bypassed. While the focus on Hughes work is often trained on the individual system builders involved in each urban context, central to his analysis is the complex of social and material entities through which both processes of transformation and urbanization were mutually constituted. That Chicago and Berlin came to be critical sites through which such transformations took place was related not only to the conjuncture of social interests and material entities that could be mobilized in each context, but also to the work involved in framing particular novelties and potential trajectories as desirable or otherwise, and the politics of this process. Here, socio-technical transformation requires the generation of spaces of authority through which multiple elements – technologies, resources, norms, beliefs – are enrolled and reassembled.

We suggest that, as a means through which governing is pursued, climate change experimentation involves just such a production of spaces of authority, creating new rationalities, techniques and subjectivities around which urban socio-materialities can be realigned and transformed. Developing this perspective, Monstadt (2009) suggests that urban infrastructure regimes are critical to the ways in which transitions unfold in urban contexts. For Monstadt (2009: 1937), urban infrastructure regimes are 'stable urban configurations of institutions, techniques, and artifacts which determine "normal" socio-technical developments in a city and thus shape general urban processes and the urban metabolism'. They are held in place through structures of power, strategic endeavour, and mundane techniques and practices, and provide the means through which infrastructure provision takes place and metabolic circulation through and beyond the urban is enabled (Bulkeley *et al.* 2013). Yet, at the same time, an urban political ecology perspective points to the emergent and unruly dynamics of such systems and circulations, such that continual work is required to maintain and normalize regimes, and the potential is always present for things to be otherwise. Furthermore, multiple regimes may coexist, each configuring relations between power, nature and urban infrastructure in distinct ways. From this perspective, making space for the urban within analyses of the potential and prospects for socio-technical transition means acknowledging both the multiple and fragmented nature of regimes, and that what is at stake in the transformation of the ways in which societal functions are met is the reconfiguration of different forms of urban metabolism, a point to which we return below.

Beyond scaling up: gaining traction

Examining how the dynamics of urban metabolisms configure distinct urban regimes and mediate power relations between different agents requires that we revisit the ways in which experiments have been conceived as providing the basis for system transformation. Within the literature on socio-technical transitions, experiments are regarded as one-off initiatives designed to test the boundaries of innovation and which can act as a fulcrum for diverse interests and ideas to be drawn together. At stake in understanding how experiments come to matter has been the analysis of the ways in which such initiatives come to form niches:

> ... defined in the literature as 'protective spaces' where real world experimentation and development of sustainable technologies can take place and supportive constituencies can be built. Niche protective spaces shield the innovation against premature rejection by incumbent regime selection pressures, until the innovation is proven to be sufficiently robust to compete and prosper in unprotected market settings.
>
> (Smith *et al.* 2013: 2)

The key dynamics at work in the stabilization of niches to provide the basis for transitions are growth (of the technology and the social networks that support it) and learning, such that actors involved in particular niches come to share common understandings about the technology and its transformative potential (Brown and Vergragt 2008). In classic innovation studies, such niches break through and are able to diffuse across time and space. Emphasis is placed on the need to scale up experiments from singular examples or particular contexts in order that they can have sufficient influence to be widely taken up. For the most part, such processes of scaling up are conceived in terms of either enlarging the experiment in any one place or its replication in different contexts, although it is also recognized that scaling up can occur when the essential elements of any particular project are transferred or the salient lessons for other actors extracted and incorporated in other institutional arenas. Across these extensive bodies of work there is much debate about how such processes of transfer and learning take place, and of the challenges of extracting and transposing examples and lessons in particular contexts. However, for the most part, experiments (or best practices, initiatives, innovation, depending on the terminology being used) are regarded as technologically and institutionally bounded and transferable, reproducing the sense that such entities are one-off exemplars, and neglecting the socio-technical work involved in their continual reproduction. Yet, as literature on policy mobilities (e.g. Temenos and McCann 2012) has shown, the transfer of lessons or innovations rarely takes place in such a coherent or directed manner. Policy lessons are not simply contained and transferred, but rather can be seen as emerging from within the assemblage of multiple, coexisting elements from which policy-makers can draw more or less opportunistically, through the enrolment of actors and materials, and the appropriation of discourses of best practice or guidelines to suit political interests within a given context (see also Chapter 1).

In contrast to focusing on technical and policy innovation as a process of diffusion, the socio-technical transitions literature has sought to understand how niches provide the basis for transformation through processes of co-evolution and alignment between innovations, regime stress, and landscape pressure (Geels 2005; Geels and Schot 2007). Rather than necessarily presupposing that experimentation provides the basis for the success or failure of niches, the multilevel perspective on systems in transition seeks to identify multiple pathways that could potentially arise as a result of the combination of different forms of alignment. For example, Geels and Raven (2006) propose four main pathways that arise from different combinations of the timing and kinds of interaction taking place across the multilevel system: transformation (which Berkhout *et al.* 2010 suggest is more aptly termed reorientation), reconfiguration, technological substitution, and de-alignment and re-alignment. However, here too new pathways depend on the conceptualization of niches as bounded entities which come to have effect when they diffuse from their origins, exploiting windows of opportunity or expanding and, in so doing, adjusting, substituting or competing in the regime until a new dominant form is achieved. For the most part, such approaches have tended to assume that niches emerge from outside existing dominant regimes and that regimes themselves are relatively stable, homogenous and encompassing. However, such assumptions are difficult to sustain in the face of the co-presence of multiple regimes and the interweaving of incumbent regime actors in the process of experimentation (Smith 2007; Bulkeley *et al.* 2013).

Recent work points to an alternative way in which the notion of pathways to transition could be used, not to identify singular transitions from one current state to another but to encourage the recognition of the wide diversity of imagined urban futures that coexist within a single city (Rydin *et al.* 2013). They argue for the need 'to think carefully and critically about how particular choices, whether social or technical, prefigure the adoption of a particular direction of travel, while potentially closing off alternative destinations' (Rydin *et al.* 2013: 637). For Rydin and colleagues (2013), a pathways approach enables an analysis of the political and contested dynamics of energy systems, and a means of identifying the multiple alternative configurations that are currently emerging in the UK. Similarly, Leach *et al.* (2010: 48) suggest that a pathways approach starts from the assumption that 'different actors and networks, framing systems dynamics, boundaries and goals in different ways, produce very different narratives about what a response should be and what might make it effective' in terms of realising sustainability. Rather than seeing niches as emerging separately from regimes, this work points to the ways in which experimentation is articulated with existing socio-technical configurations. Other scholars have sought to examine this dynamic more closely through the concept of translation (Smith 2007; Seyfang 2010) that occurs: (1) as a result of *problematization*, where problems within regimes open up the possibilities for niches, either in terms of new problem framings or financial incentives, for example; (2) in the form of *integration*, where niches come to more closely resemble the regime through developing 'add on' technologies or incremental adjustments to efficiency standards, e.g. in the case of low carbon housing; or (3) as a result of *intermediation*, through the development of projects as partnerships between niche/mainstream actors which provide spaces for niche ideas to be taken up by regime incumbents. Developing this line of analysis, Smith *et al.* (2013) have sought to explore how niches operate as a space through which system transformation can emerge, pointing to their important role not only in shielding and nurturing innovation, but as spaces of empowerment. Empowerment is regarded as taking two distinct forms, one in which the niche innovation is able 'to fit and conform' into incumbent regimes (Smith *et al.* 2013: 3), and the second in which it is able 'to stretch and transform' them. In this way, '[m]easures constituting niche protective space become institutionalized into a reformed regime, including re-structured markets' (Smith *et al.* 2013: 4). From our perspective, empowerment could also be engaged in a more traditional sense in order to examine how experiments come to matter, through challenging hegemonic understandings and developing alternative forms of socio-environmental politics.

Rather than viewing niches and regimes as separate entities, such that the effectiveness of niches is dependent on the degree to which they can be scaled up or gain sufficient strength to create a transition from one steady state to another, emphasis on translation and empowerment hint at the ways in which experimentation and socio-technical systems are mutually and relationally constituted. Attending to the ways in which experiments garner transformations in socio-technical and socio-ecological systems therefore requires an engagement with the ways in which they are elaborated, sustained and come to be mobilized. Rather than being a matter of scaling up experimentation, its transformative potential comes through the traction generated by refashioning existing socio-technical configurations by translating between climate governmentalities and existing forms of social and material order. Analysing the case of photovoltaic systems (PV) in the UK, Smith and colleagues (2013: 12–13) find that 'fragmented and sometimes fleeting spaces have emerged' together with 'measures whose realisation over time addressed particular and immediate aspects of PV, and that were returned to and picked up when later measures provided further occasions for development'. Rather than providing the basis for transition, experimentation can be regarded as creating the sites and moments through which new configurations are assembled, challenged and reworked. For Smith *et al.* (2013: 4), this requires analysis of 'the negotiation, bargaining and compromises over spaces for sustainable innovation'. We seek to extend this approach by arguing that it is not only the discursive form through which such experimentation is constructed and contested that matters, but rather the socio-material dynamics of governing experimentation through making, maintaining and living that are at work in shaping the possibilities and limits of transforming socio-technical systems.

Topological shifts: the transformative potential of making, maintaining and living experimentation

Rather than regarding transitions as the wholesale replacement of one socio-technical regime with another, we argue that transformation involves purposive attempts to reconfigure socio-technical and socio-ecological systems in relation to climate change. Here, the urban is neither an actor nor an arena within which such transformations play out, but co-constitutive of both the regimes and experiments through which these processes unfold. Instead of placing emphasis on the means by which interventions are able to effect a transition through scaling up or developing new pathways, we suggest that we need to attend to the ways in which experimentation serves to shift the topologies of socio-technical and socio-ecological systems, gathering new forms of agency and disrupting the obduracy of urban systems. Such shape-shifting requires both that sufficient transformative potential is generated through experimentation and that there is some malleability within the systems of which it is a part. Reconfiguring topological landscapes in turn draws new elements into relation with one another and affords new forms of circulation, while serving to distance and marginalize other urban constituents. Within socio-ecological and socio-technical networks, an experiment acts as source of disturbance which is not confined to a particular place and moment, but rather reverberates through the network with the potential to reconfigure its constituent elements and the circulations of which they are part. An illustration might be found in that of a boiling viscous liquid, such as the mud pools found in Rotorua, New Zealand, where bubbles of air not only disturb the surface but create new patterns, imprints of their passing, which in turn shift the surface topology, creating new forms of potential. Such a reading of the transformative potential of experiments suggests that this rests on the ways in which experiments are mobilized and gain momentum, affording new topological shifts, such that metabolic circulations might be adjusted, accommodated or re-routed.

As we have argued throughout this book, central to the dynamics of experiments are the processes through which they are made, maintained and lived. In the *making* of experimentation, we found processes of assemblage which work together to configure the experiment and make it visible, to position it in relation to existing infrastructure regimes, and to establish its spatial and temporal reach. Exploring the ways in which these processes become central to their transformative potential, we find that three facets are critical. First, as a practice of assemblage, the making of experiments involves the 'collage, composition and gathering' (McFarlane 2011) of socio-material elements drawn from a range of agents operating across multiple temporal and spatial scales. This in turn requires the folding together of elements such that entities that appear at a distance from the urban milieu are gathered and drawn together in particular experimental sites, while others which are more proximate are held at bay (Allen 2003). Through this process of assembling, diverse spatial scales are already built into experiments, such that their potential for circulation is built into the very process of experimentation – for example, climate finance moves in different circuits to those garnered through the donations of local businesses, and such spatialities and flows are encoded into experiments from the outset. Second, making experiments involves the introduction of what Furlong (2011: 463) terms mediating technologies which create 'small changes to the peripheral nodes of an infrastructural network ... that leave the core of the system intact', but 'bring about important changes' through recasting the visibility of networks, reworking the relationships between users and providers, and rearticulating what infrastructure networks are for. In such practices of making experiments, the entity itself – whether it be a housing development or a refurbished roof – acquires different momentum, in turn intersecting with, gathering and transforming the landscape through which it travels in relation to the will to improve climate change. Reconfiguration also takes place through the techniques of calculation deployed in making experiments, such that the value or purpose of particular entities is reassessed in relation to new problematizations. Smith *et al.* (2013: 11) argue that the development of European Union demonstration programmes served to create new momentum for PV in the UK by reframing PV not as a form of stand-alone power generation but as building an integrated technology, with the result that 'expectations shifted, and the basis for assessment could be opened to new criteria'. Reconfiguring PV in this manner opened up new routes for the technology to circulate and become embedded within socio-technical systems.

If the *making* of the experiment enables us to examine how such interventions might lead to the in situ reconfiguration of the spatialities and materialities of socio-technical networks, enabling the mobilization and circulation of climate governmentalities, the *maintenance* of experiments, suggests that the ways in which such experiments intersect with and are related to multiple forms of urban metabolism are critical in its transformative potential. Experiments intermediate in urban flows through forms of metabolic adjustment. The extent to which experiments may be successful in this endeavour depends upon what Roy (2012: 33) has termed 'circulatory capacity, how worlds are put into motion at and through such nodes'. Roy suggests that such circulatory capacity can readily be found in well-established 'centres of calculation', such as the World Bank, but that 'counterforces of development organizations like the Grameen Bank [also] seek to exert similar types of circulatory capacity' (Roy 2012: 39). The circulatory capacities of experiments, we suggest, relates to the extent to which they have been able to embed and redirect circulations in relation to climate governmentalities.

The *living* of experiments is also essential to the transformative potential of experimentation. Experiments serve to reconstitute what is involved in the conduct of various forms of everyday practice and the extent to which experiments circulate is in turn dependent on the ways in which these practices come to be reproduced. Theories of social practice point to the importance of this dynamic in structuring not only the social world, but in reproducing socio-technical arrangements and systems of provision (Spaargaren 2011). Experiments may circulate through the continual reproduction of the practices of which they have become a part. This process is inherently emergent, fragile, and open to continual contestation and reinterpretation, yet it is also constitutive of the ways in which systems of provision come to be structured, such that new forms of practice may in turn shape the ways in which urban services are produced (for example, the forms of housing or transportation that are required in order to practice certain kinds of low carbon living). Further, climate change experimentation entails the creation, uptake and resistance of new forms of subjectivity which seek to relate what constitutes normal and appropriate conduct in response to climate change. Through the creation of new forms of low carbon or resilient climate subjects, experiments seek to create new expectations about what is (and is not) normal. The extent to which such norms may circulate is dependent on the extent to which they may relate to, and consciously be joined with, the development of such norms taking place at other sites and scales, explicit attempts to spread the message beyond the experimental domain through communication and engagement strategies aimed both at other governmental actors and publics, and demonstration of what living such norms entails in practice.

Viewing transformation not as transition in its linear guise, but rather in terms of the reconfiguration and reordering of socio-technical and socio-ecological networks, the analytical imperative becomes not one of examining how experiments might open up new pathways or be scaled up to replace existing regimes, but of understanding how and why they might gather sufficient potential and traction to *shape-shift* the regimes of which they are a part, creating new forms of circulation and relation across the urban landscape. We argue that this potential is generated through the processes and practices by which experimentation is made, maintained and lived. In practice, the ways in which experimental potential is translated and empowered depends both on its particular characteristics and the malleability of the topological networks within which they are situated, and it is to these issues that we now turn through an analysis of the cases included in this book.

Transformation and mobilization

The mobilization of experiments depends, we argue, not on external forces of diffusion, transfer or scaling up, but rather on the potential garnered through the processes of making, maintaining and living that are inherent to experimentation. Thus, the transformative potential of experiments relates to the enfolding of socio-spatial relations of power within experiments, the ways in which they reconfigure existing assemblages to entrain different socio-materialities and ecologies, their circulatory capacities in relation to wider circuits of urban metabolism, the extent to which they give rise to new forms of practice and its reproduction, and the ways in which they establish new norms in relation to climate governmentalities. The extent to which experimentation produces forms of transformation relies on the nature and extent of these processes, and how they serve – singularly and collectively – to shift the topological networks of which they are a part: translating existing metabolic circulations onto different co-ordinates, opening up new circuits through which to sustain low carbon or resilient forms of urban socio-technical systems, and empowering new political dispositions.

Table 12.1 compares the extent to which the experiments discussed in this book embody transformative potential in relation to the ways in which they are made, maintained and lived. We are mindful of the difficulties of developing a comparative analysis of what are, in essence, different types of interventions, relating to diverse forms of climate governmentality, and occurring in multiple contexts. Rather than regarding experiments as directly comparable, here we seek to draw out what each might tell us about particular urban instances of the processes of making, maintaining and living, and how these collectively contributed to establishing different forms of momentum in each case.

As set out in Table 12.1, each experiment is able to create a shift in the urban fabric through the processes of being made, maintained and lived, though the momentum and traction that is created in each case varies in important ways. Beyond the dynamics of experimentation itself, recognising that the transformative potential of experimentation is realized through these forms of mobilization also entails attending to the manifold sites and moments that may be caught up in the shifts that experimentation gives rise to, as well as the multiple and emergent forms that such shifts might create. This multiplicity can be regarded as constituting the urban milieu (discussed in Chapter 2) which Foucault described as 'the space in which a series of uncertain elements unfold' and regarded as necessary in order to

TABLE 12.1 The transformative potential of experiments

Experiment	*Making*	*Maintaining*	*Living*
T-Zed Bangalore	Multiple mediating technologies rearticulate what housing can become	Securitisation of resources and integration of innovation into housing economy	Constitution of the middle-class green consumer
ViDA Monterrey	Calculative practices mediate new configuration of housing and finance; draw together actors operating at different scales	Lack of maintenance and unsuccessful attempts to re-experiment	Experiment is appropriated within conventional living practices and norms
Coolest Block Contest Philadelphia	Forges public-private-community partnership; tested technology made newly visible, revealing (economic) potential of roofs	Established to minimise repair; circulation limited to particular parts of the urban landscape	The creation of low carbon subjects at an individual level, but their enactment in collective activities
Retrofitting Mamre Cape Town	Simple mediating technology and carbon-based financial calculation draw together local/international networks	Active articulation with metabolic flows of health, economy, urban energy security and international development agenda	Generation of new norms that link low carbon with dignified housing within communities and elites
Solar Atlas Berlin	Narrow calculative approach confined to a small number of actors	Lack of maintenance and embedding in the solar economy; becomes a transitory moment	The experiment is not lived beyond its assemblage
SHW Sao Paulo	SHW mediates multiple urban socio-technical and ecological networks, gathering momentum through array of actors and calculations	Strong effort to maintain the experiment both in terms of upkeep and repair but also circulatory capacity within wider metabolisms of energy and water	Creation of low carbon subjects through norms of dignified housing and integration of innovations within everyday practice which are reproduced
Climateers Hong Kong	Forges public-private-community partnership; new calculative devices enrol public participation	Maintenance of circulatory capacity through social networks, more effective than education programs	Organic normalisation of low carbon subjectivities through the progressive enrolment of citizens in the experiment
Brixton LCZ London	Creates a crucible of multiple agendas together with calculation methods that lead to persuasion and enrolment	Combination of efforts to make the experiment visible and enrol the community as a key driver agent	Integration with other parallel agendas leads to its forms of low carbon subjectivity that accord with well-being

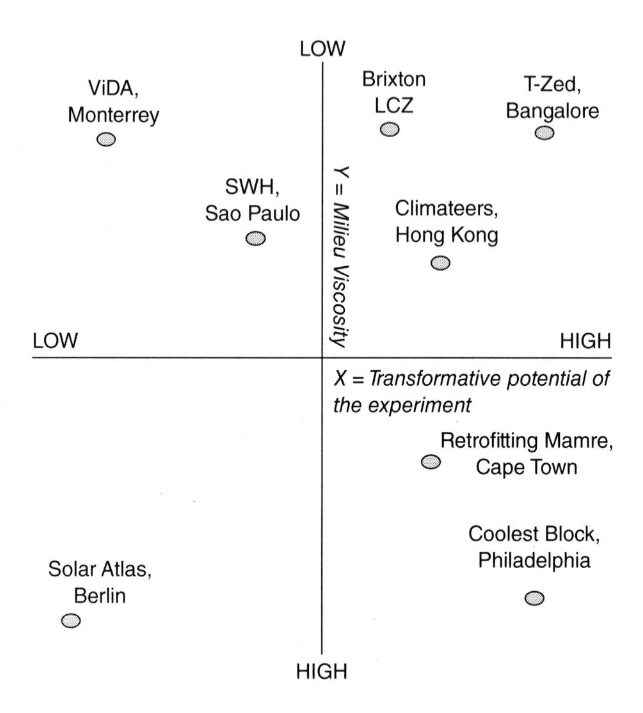

FIGURE 12.1 Transformative potential and milieu viscosity for each experiment.

'account for action at a distance of one body on another. It is therefore the medium of an action and the element in which it circulates' (Foucault 2009: 21). From this basis, the mobilization of urban climate change experiments can be conceived both in terms of their transformative potential, but also in relation to the obduracy or viscosity of the milieu in which they unfold and circulate. Figure 12.1 provides an overview of the different forms of mobilization produced through the experimentation in relation to these two different axes – transformative potential and milieu viscosity.

One experiment that scores highly on both axes is T-Zed, Bangalore. In T-Zed transformative potential was driven through the serial accumulation of innovations which served to rearticulate the nature of low carbon housing and to normalize the notion of a low carbon consumer for the middle classes. Yet such forms of subjectivity and their implications for everyday practice remained contested and, even in a context where resource security concerns are paramount, the capacity of the experiment to adjust metabolic flows beyond its own borders was relatively weak. Nonetheless, the experiment gained considerable momentum through its insertion in Bangalore's housing and resource economy and, through the twin emphases on resource security and low carbon development, was able to generate considerable malleability across the urban landscape, in which accommodating development alongside environment/resources concerns is a pressing political concern and where an emergent middle class are willing to buy into new norms of low carbon living.

T-Zed was able to create a shift in the housing infrastructure networks of which it was a part and, as a result economies, technologies, norms and practices established within T-Zed have started to circulate across the city and beyond.

An experiment that draws on the potential for change from the urban milieu, rather than from its transformative potential, was ViDA, Monterrey. ViDA focused on sustainable housing design innovations and their integration within the dominant templates of the housing industry. The project succeeded in bringing an array of national and local actors together to constitute a new arena for intervention, but also in making innovation invisible and cost free within the dominant housing regime. At the same time, the lack of maintenance and the minimum reconfiguration of living practices involved have led the project to be normalized and integrated with the landscape of housing, through the localized re-assemblage of its social and material components. Together, these findings suggest that the experiment had low transformative potential. However, the housing regime of which it was a part sought to adopt those elements which accorded with securing housing finance, where the installation of appliances would increase the value of the houses, and of the mortgages that residents should pay, but left behind the more challenging principles which required a careful consideration of the locale and which resisted replication in design terms, such as orientation. Through contributing to the development of the Green Mortgage scheme, ViDA has served to work through a specific urban milieu to re-orientate the housing regime, shifting its topology and through this creating new forms of green economic circulation and reductions in the flows of greenhouse gas (GHG) emissions from new housing development. For many, including those who have recognized the Green Mortgage scheme as internationally leading, this has been heralded as transforming the housing landscape. Yet the housing regime has remained relatively unchanged in its relation to the challenges of climate change, sustaining a strong growth logic while, at the same time, the extent to which green mortgages are leading to the maintenance of housing configurations and everyday practices that can reduce GHG emissions in the long term remains moot.

In the Coolest Block contest, Philadelphia, we find significant transformative potential, but a relatively viscous or obdurate urban milieu which served to constrain its circulation. The experiment demonstrated the possibilities of public–private partnerships for retrofitting housing and how, in particular, they could enrol the public to support interventions. Here, a tried and tested technology was made newly visible in the city, while a hidden resource – the roof – also came to be subject to intervention. Yet, despite the efforts in articulating and mobilizing these partners, technologies and resources, the transformative potential of the experiment was reduced by working through established modes of engagement, which served to train its focus on individual champions as the drivers of change and the momentary nature of the competition. While the experiment did serve to make the need for cool roofs visible in the city, influencing the development of local development policies, it remained marginal to the housing and energy regimes, which were relatively unchanged by its presence. Furthermore, in the complex socio-material context of poor housing and limited resources in Philadelphia, the experiment was perhaps least able to engage with those parts of the urban fabric that needed it most, but which remained outside its orbit.

The Solar Atlas, Berlin, is the experiment in which we see least evidence of transformation, both in terms of the potential of the experiment and its translation through the urban milieu. Here, the focus was in aligning economic and technical criteria through assembling the instruments that would make the solar potential of Berlin visible and thus amenable to investment. The social aspects of the experiment, from how it would be publicly adopted and the institutional and policy arrangements that would make it possible, were largely overlooked. The emphasis was very much directed towards technological development, but there was no consideration of the difficulties that maintaining that experiment over time would entail. Thus, the way the experiment was assembled made it impossible to either maintain it (a set of processes to which insufficient resources were allocated) or to be lived (which was never considered). In Berlin, the strategic calculations of the intervening actors appeared to overlook considerations about the particular moment and place of intervention, or the means that would be required to shift the socio-technical configurations of housing and energy provision in the city. The experiment was hardly implemented and it has largely failed to have any imprint in the economies or lives of Berliners. While the initial process of problematization may have helped to open up the field of experimentation to other solutions trained on creating new forms of energy resource, it is fair to say that the Solar Atlas experiment has had little influence on the governance of climate change in Berlin, although it may have had influence in knowledge development fields beyond Berlin (e.g. MIT Solar Atlas) which pick up on the idea rather than on the practice or outcomes of the experiment.

In the other four experiments examined in this book, the combinations of transformative potential and urban vicosity are more ambiguous. In Mamre, Cape Town, housing retrofit was enabled through a mode of calculation of the potential for carbon offsetting in low-income houses first conducted in Kuyasa, and the ability of relatively low-cost and readily-available technologies to mediate between international flows of climate finance and the needs of housing provision. Advancing the experiment in Mamre required that the logics of climate finance become entrained with community needs – for better standards of living, economic possibilities and improved health – and the experiment successfully managed to adjust these circulations in relation to climate finance. In a context where international networks and development finance are critical to urban politics, the experiment was readily inserted, providing rationales and techniques which have continued to reverberate due to their potential for accessing finance for realising housing and climate change imperatives within the energy and housing regimes of the city. At the same time, such interventions require specific alignments of finance and infrastructure in order to be assembled and mobilized, and it is clear that access to such resources is limited across the city, so that its potential to offer a transformative model for either housing or energy provision has been limited by the existing structures of inequality and resource access that pervade the urban milieu.

In Sao Paulo, the emphasis was on the integration of the institutional and material aspects that would make the installation of solar hot water (SHW) both possible and compelling. Even though SHW is a relatively well-known technology, its potential was considered anew in the light of the multiple agendas of low carbon development, energy security and the challenges of peak demand. Once it came to be considered as a potential solution to the climate–energy problem in the city, technologies had to be calculated and made compelling within existing protocols and working practices, such that it could be made to fit within the system of providing and building social housing. The work of maintaining the experiment has led to further reconfigurations of SHW and the ways in which it can be made to fit into the urban landscape, requiring the disturbance of established notions of housing as a basic need with the idea of dignified and sustainable housing provision. The insertion, upkeep and adoption within the life practices of residents of the technology in different urban contexts and its circulation across the housing/energy nexus enabled SHW to gain traction. At the same time, its salience for multiple actors and governmental problems enabled it to reform the urban milieu within which it was situated, particularly through drawing electricity utilities into the arena of housing provision and through its take up within the 'Minha Casa, Minha Vida' programme which have served to create new economies through which SHW can circulate and gain momentum. At the same time, the provision of SHW is itself rather limited in terms of its transformative potential for achieving low carbon transitions, given that it is focused on only one form of energy use and the specific practices associated with the use of hot water. Its wider implications in terms of household consumption or the provision of other renewable resources appear to be rather limited.

In Hong Kong, the Climateers experiment successfully drew into conjunction corporate and environmental organizations concerned to develop a response to climate change. The materialization of the processes of carbon calculation – and their subsequent maintenance – was central in both forging and sustaining the alliances that made the experiment possible, and in engaging individuals in the governing of carbon across the city through creating new subjectivities and practices for the low carbon citizen. However, the case illustrates that, while new discursive norms readily circulated, ensuring their material manifestation in a context hard-wired for energy consumption was challenging and contested. High carbon consumption is built into the city, both physically and culturally, such that the potential for the forms of climate self-government engendered through Climateers to shift this topology were relatively weak. The Brixton LCZ, London, represents an experiment in which enrolling relevant actors has meant a transformation of the experiment itself, from an initial rhetoric of narrow targets and the implementation of a specific set of measures, to a more organic deployment of the experiment in relation to community concerns and ongoing activities in Brixton. This flexibility has enabled the experiment to work through the urban milieu and become inscribed in the landscape of local climate change governance, which in turn has served as a way to ensure its maintenance and appropriation within living practices over time. While in the case of Monterrey a similar process led to the virtual merging of the experiment with its milieu – to the extent to which its authorization as an instrument for climate change governance was lost – in Brixton the opposite was true. Here, it was only by bringing the experiment in line with community expectations and practices that it has gathered traction. Thus, this example shows how the transformative potential of the experiment may change through its interaction with the milieu.

Overall, the examples explored here demonstrate that the transformative potential of the experiment is already embedded in the processes that lead to it being made, maintained and lived; however, being successfully accomplished is not a guarantee that this transformative potential will be realized. For example, the Solar Atlas was perhaps the most technologically innovative, successfully-assembled experiment we considered, but its transformative potential was hindered by its lack of maintenance and limited attempt to integrate it in living practices. In contrast, the Climateers project was perhaps not so successfully assembled, and yet it was, by the continual reproduction of the process of making the experiment, the maintenance of appropriate circulations and an active process of subjectification, that the programme has led to arguably important changes in relation to how climate change action is perceived in Hong Kong. Projects, such as ViDA in Monterrey or the Coolest Block contest in Philadelphia, demonstrate in contrast that the transformative potential of any experiment can only be made apparent in relation to the obduracy or viscosity of the urban milieu in which it is situated. The potential for experiments to take effect, become circulated through ever wider networks, establish new forms of economy, or gain momentum within the practices of households, businesses and political arenas is thus forged in the relation between the processes that sustain its potential – the making, maintaining and living of experiments – and the ways in which the urban milieu is constituted and challenged.

Conclusions: re-engaging the urban politics of climate change

> As a *milieu of liberal government* the city becomes a sort of laboratory of conduct. Its government comes to be seen as essentially problematic, so the city becomes a plane of indetermination – a dense, opaque, unknown, perhaps ultimately unknowable place; a domain where the criteria and techniques of good government were no longer self-evident.
>
> *(Osborne and Rose 1999: 740, emphasis in the original)*

Experimentation is a mode of governing climate change that now pervades multiple cities around the world. Through surveying the landscape of climate change experiments (Chapter 1), developing a conceptual account of how governing by experiment is accomplished (Chapter 2), and the analysis of eight case-studies from global cities (Chapters 3–10), we have sought to develop an account of this phenomenon and its implications for how we engage with the art of governing climate change in the city. We have argued that their consequences for social and environmental justice (Chapter 11) as well their transformative potential (this chapter) are not created by the specific agents, institutions, technologies or social innovations that made them possible, but is instead produced as experiments are configured within a given context through the three-fold processes that we have termed making, maintaining and living. As Osborne and Rose (1999) suggest, the problematization of the urban milieu as a space within which governmental intervention is required has historically been bound to questions concerning what good government might entail and how it should be achieved. As the urban comes to form a new laboratory for conduct regarding climate change, this insight can help us to reflect, in conclusion, on the nature of experimentation and its politics.

First, we suggest that the emergence of experimentation does indeed reflect a pervasive sense of indeterminacy about how responses to climate change should take place. Confronted not only with scientific uncertainty about the effects of the changing environment in particular places or which actions might be most effective in mitigating these conditions, but also with unknown unknowns concerning the unfolding processes of urbanization and the viability of different forms of innovation, experimentation has become a means through which a response can be articulated. As a conscious process of seeking to try out, innovate, and put to the test various combinations of more or less known technologies, techniques and forms of organization, experimentation serves as a means through which intervention can proceed despite the opacity of the domain into which they are inserted. Here, we have argued, the key task of making experiments serves to create more visibility and certainty, delimiting the field of intervention, creating specific alignments that pull climate change into more well-worn tracks, applying calculative techniques that create 'known knowns', and producing reasons to engage. Maintaining experiments in turn creates momentum, both in the incremental process of upkeep and through attempts at adjusting metabolic circulations to account for and enable climate responses, while the processes of living experiments create the new forms of subjectivity that are capable of undertaking low carbon conduct and reframing everyday practice in relation to new norms. Through these processes, particular urban milieu come to be understood in relation to climate change, and what constitutes good government, and the improvement of the urban population, seen through this lens.

Second, as good urban government comes to be concerned with climate change, so too does it serve to relate existing realities to the notion of what the good city could become. As Anderson has argued:

> in the enactment of better worlds, the future is constantly being folded into the here and now; a desired future may act as a spur to action in the present, for example, or action in the present may bring back memories of long-forgotten hoped-for futures.
>
> *(Anderson 2010: 778)*

Rather than taking place outside of existing regimes or political economies, experimentation emerges from within them, translating their potential and their problems into other sites which appear to hold the promise of accomplishing new kinds of urban future. As Alice found when she returned to the river bank after her adventures in Wonderland, such dreams are after all often refractions of existing realities. For Foucault, such sites are heterotopias, spaces 'that have the curious property of being in relation with all the other sites, but in such a way as to suspect, neutralize, or invert the set of relations that they happen to designate, mirror, or reflect' (Foucault 1986: 24). Experiments in this sense are neither the nowhere of utopian visions, nor do they simply reproduce the conditions of existing regimes and systems. Rather they function as a form of other space 'in which new ways of experimenting with ordering society are tried out' (Hetherington 1997: 12). Rather than being generated through their displacement from one site to another, we argue that the potential for experiments to achieve transformation is situated and generated through the forms of socio-material relations they generate, and in relation to the obduracy of the urban milieu of which they are a part.

Finally, in drawing attention to experimentation as an art of government, we also seek to excavate its politics. Where the authority to govern climate change is dispersed, the will to improve has to be generated and negotiated across and between multiple agents. Achieving the right disposition of things, the socio-technical configurations through which government is conducted and new forms of disposition and practice realized requires the ongoing work of making, maintaining and living experiments. Such interventions shift existing social and spatial relations, folding entities into new relations, such that, for example, the evening shower becomes a matter of low carbon energy provision in Sao Paulo, or structures of housing finance are rerouted through suburban development in Monterrey. Such shifts and folds serve to both gather and include, and make distant and exclude. In so doing, experimentation can constitute an urban politics that can be engaged for progressive ends, as governing climate change comes to be regarded as a means of achieving other forms of social, economic and environmental benefits for those who have previously been excluded. Yet it can also serve to sustain existing forms of inequality, reproducing divisions or fostering new forms of exclusion on the basis of access to climate benefits or exposure to risk. Their outcomes are uncertain, and the intentions multiple – from realising new forms of carbon control or ecological security, to fostering empowerment or more flexible urban spaces. The political project of experimentation can therefore only be partially designed, and remains always subject to ongoing contestation and to the reconfiguration of circuits of accumulation and dispossession for 'the city, as domain of immanence … remains an open-ended provocation to government' (Osborne and Rose 1999: 759). Rather than serving as a means through which politics is evacuated, what the art of experimentation tells us, above all, is how and by whom rights, responsibilities and recognition in relation to the urban politics of climate change are manifest. And ultimately what this means about the kind of society we want to become.

References

Allen, J. (2003) *Lost Geographies of Power*. Oxford: Blackwell.

Anderson, B. (2010) 'Preemption, precaution, preparedness: Anticipatory action and future geographies'. *Progress in Human Geography*, 34(6): 777–98.

Bailey, I., Hopkins, R. and Wilson, G. (2010) 'Some things old, some things new: The spatial representations and politics of change of the peak oil relocalisation movement'. *Geoforum*, 41(4): 595–605.

Berkhout, F., Verbong, G., Wieczorek, A. J., Raven, R., Lebel, L. and Xuemei Bai (2010) 'Sustainability experiments in Asia: Innovations shaping alternative development pathways?'. *Environmental Science and Policy*, 13: 261–71.

Brown, H. S. and Vergragt, P. J. (2008) 'Bounded socio-technical experiments as agents of systemic change: The case of a zero-energy residential building'. *Technological Forecasting and Social Change*, 75(1): 107–30.

Bulkeley, H., Castán Broto, V. and Maassen, A. (2014) 'Low-carbon transitions and the reconfiguration of urban infrastructure'. *Urban Studies*, 51(7): 1471–86.

Carvalho, L., Mingardo, G. and Van Haaren, J. (2012) 'Green urban transport policies and cleantech innovations: Evidence from Curitiba, Göteborg and Hamburg'. *European Planning Studies*, 20(3): 375–96.

Coenen, L., Benneworth, P. and Truffer, B. (2012) 'Toward a spatial perspective on sustainability transitions'. *Research Policy*, 41(6): 968–79.

DECC (2009) *The UK Low Carbon Transition Plan*. London: DECC.

Elzen, B. and Wieczorek, A. (2005) 'Transitions towards sustainability through system innovation'. *Technological Forecasting and Social Change*, 72(6): 651–61.

Foucault, M. (1986) 'Of Other Spaces'. Trans. J. Miskowiec. *Diacritics*, 16(1): 22–7.

——(2009) *Security, Territory, Population: Lectures at the College de France, 1977–78*. Ed. M. Senellart, trans. G. Burchell. Basingstoke: Palgrave Macmillan.

Furlong, K. (2011) 'Small technologies, big change: Rethinking infrastructure through STS and geography'. *Progress in Human Geography*, 35(4): 460–82.

Geels, F. (2002) 'Technological transitions as evolutionary reconfiguration processes: A multi-level perspective and a case-study'. *Research Policy*, 31(8–9): 1257–74.

——(2005) *Technological Transitions and System Innovations: A Co-evolutionary and Socio-technical Analysis*. London: Edward Elgar.

——(2010) 'The role of cities in technological transitions'. In Bulkeley, H., Castán Broto, V., Modson, M. and Marvin, S. (eds) *Cities and Low Carbon Transitions*. London: Routledge, pp. 13–28.

Geels, F. and Raven, R. (2006) 'Non-linearity and expectations in niche-development trajectories: Ups and downs in Dutch biogas development (1973–2003)'. *Technology Analysis and Strategic Management*, 18: 375–92.

Geels, F. and Schot, J. (2007) 'Typology of sociotechnical transition pathways'. *Research Policy*, 36(3): 399–417.

Graham, S. and Healey, P. (1999) 'Relational concepts of space and place: Issues for planning theory and practice'. *European Planning Studies*, 7(5): 623–46.

Hetherington, K. (1997) *The badlands of modernity: Heterotopia and social ordering*. London: Routledge.

Hodson, M. and Marvin, S. (2010) 'Can cities shape socio-technical transitions and how would we know if they were?'. *Research Policy*, 39(4): 477–85.

——(2012) 'Mediating low-carbon urban transitions? Forms of organization, knowledge and action'. *European Planning Studies*, 20(3): 421–39.

Hommels, A. (2005) 'Studying obduracy in the city: Toward a productive fusion between technology studies and urban studies'. *Science, Technology, & Human Values*, 30(3): 323–51.

——(2008) *Unbuilding Cities: Obduracy in Urban Socio-Technical Change*. Cambridge, MA: The MIT Press.

Hubbard, P. (2006) *City*. Abingdon, Oxon: Routledge.

Hughes, T. (1983) *Networks of Power Electrification in Western Society, 1880–1930*. Baltimore, MD: Johns Hopkins University Press.

Leach, M., Scoones, I. and Stirling, A. (2010) *Dynamic Sustainabilities: Technology, Environment, Social Justice*. London: Earthscan.

Loorbach, D. and Rotmans, J. (2010) 'The practice of transition management: Examples and lessons from four distinct cases'. *Futures*, 42(3): 237–46.

McFarlane, C. (2011) *Learning the City: Knowledge and Translocal Assemblage*. Chichester: Wiley-Blackwell.

Monstadt, J. (2009) 'Conceptualizing the political ecology of urban infrastructures: Insights from technology and urban studies'. *Environment and Planning A*, 41(8): 1924–42.

Mumford, L. (1997 [1937]) 'What is a City?'. In LeGates, R.T. and Stout, F. (eds), *The City Reader*. London: Routledge, pp. 183–8.

Osborne, T. and Rose, N. (1999) 'Governing cities: Notes on the spatialisation of virtue'. *Environment and Planning D: Society and Space*, 17(6): 737–60.

Raven, R., van den Bosch, S. and Weterings, R. (2009) 'Transitions and strategic niche management: Towards a competence kit for practitioners'. *International Journal of Technology Management*, 51(1): 57–74.

Roy, A. (2012) 'Ethnographic circulations: Space–time relations in the worlds of poverty management'. *Environment and Planning A*, 44(1): 31–41.

Rydin, Y., Turcu, C., Guy, S. and Austin, P. (2013) 'Mapping the coevolution of urban energy systems: pathways of change'. *Environment and Planning A*, 45(3): 634–49.

Seyfang, G. (2010) 'Community action for sustainable housing: Building a low-carbon future'. *Energy Policy*, 38(12): 7624–33.

Smith, A. (2007) 'Translating sustainabilities between green niches and socio-technical regimes'. *Technology Analysis and Strategic Management*, 19: 427–50.

Smith, A., Stirling, A. and Berkhout, F. (2005) 'The governance of sustainable socio-technical transitions'. *Research Policy*, 34(10): 1491–1510.

Smith, A., Kern, F., Raven, R. and Verhees, B. (2013) 'Spaces for sustainable innovation: Solar photovoltaic electricity in the UK'. *Technological Forecasting and Social Change*, 81: 115–30.

Spaargaren, G. (2011) 'Theories of practices: Agency, technology, and culture: Exploring the relevance of practice theories for the governance of sustainable consumption practices in the new world-order'. *Global Environmental Change*, 21(3): 813–22.

Späth, P. and Rohracher, H. (2012) 'Local demonstrations for global transitions—Dynamics across governance levels fostering socio-technical regime change towards sustainability'. *European Planning Studies*, 20(3): 461–79.

Temenos, C. and McCann, E. (2012) 'The local politics of policy mobility: Learning, persuasion, and the production of a municipal sustainability fix'. *Environment and Planning A*, 44(6): 1389–1406.

Truffer, B. and Coenen, L. (2012) 'Environmental innovation and sustainability transitions in regional studies'. *Regional Studies*, 46(1): 1–21.

Verbong, G., Geels, F. and Raven, R. (2008) 'Multi-niche analysis of dynamics and policies in Dutch renewable energy innovation journeys (1970–2006): Hype-cycles, closed networks and technology-focused learning'. *Technology Analysis & Strategic Management*, 20(5): 555–73.

INDEX

Note: **bold** entries refer to figures and tables.